# Technology Trends in
# MATERIALS
# MANAGEMENT

# Technology Trends in MATERIALS MANAGEMENT

A SYSTEMATIC REVIEW OF THE USE OF TECHNOLOGY
IN MATERIALS MANAGEMENT TO ACHIEVE COMPETITIVE ADVANTAGES

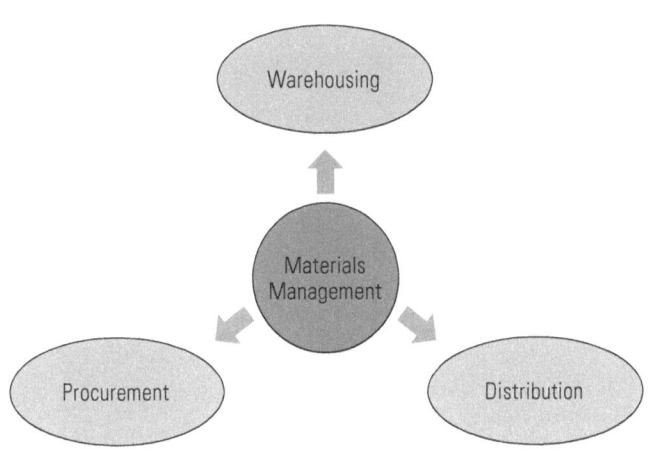

## DR. ALI KAMALI

Copyright © 2019 by Dr. Ali Kamali.

ISBN:      Softcover         978-1-7960-5236-7
           eBook             978-1-7960-5235-0

All rights reserved. No part of this book may be reproduced or transmitted in any form or by any means, electronic or mechanical, including photocopying, recording, or by any information storage and retrieval system, without permission in writing from the copyright owner.

The views expressed in this work are solely those of the author and do not necessarily reflect the views of the publisher, and the publisher hereby disclaims any responsibility for them.

Any people depicted in stock imagery provided by Getty Images are models, and such images are being used for illustrative purposes only.
Certain stock imagery © Getty Images.

Print information available on the last page.

Rev. date: 08/13/2019

**To order additional copies of this book, contact:**
Xlibris
1-888-795-4274
www.Xlibris.com
Orders@Xlibris.com
801019

# CONTENTS

List of Figures ................................................................................ix
Inspirational Words........................................................................xi
Preface ........................................................................................xiii

## Part 1: Principles and Definitions

Chapter 1:   Business Principles............................................................1
                Introduction
                Material Stock /Inventory
                The Stock Control
                Stock Planning
                Stock Records
                Materials Handling
                The Warehouse Manager
                The Procurement Manager
                The Distribution Manager
                Definition of Organisation
                Definition of Information Technology (IT)

Chapter 2:   Logistics ......................................................................18
                Introduction
                Logistics within the Supply Chain Management
                Designation of Logistics
                Task of Logistics
                Integrated Logistics Platform
                Logistics and Materials Management (MM)
                Logistics and Distribution Management
                Logistics and Procurement
                Logistics in Organisational Structures

Chapter 3:   Technology..................................................................31
                Introduction
                Initial Stage of Technology Adoption
                Technology Development Life Cycle (TDLC)

Technology Adoption Life Cycle (TALC)
Technology Acceptance Model (TAM)

## Part 2: Components of Material Management

Chapter 4:   The Warehouse .......................................................... 47
                    Introduction
                    Scope of Activities in a Warehouse
                    Functions of Warehousing
                    Warehouse Layout
                    Leasing vs. Buying a Warehouse
                    Stock Control Techniques in Warehousing
                    Material Handling Equipment (MHE) in a Warehouse

Chapter 5:   Procurement (Purchasing) ............................................ 74
                    Intruduction
                    The Importance of Procurement from MM Aspects
                    The Procurement Objectives
                    Procurement within the Organizational Structure
                    Main duties of Procurement
                    Important Aspects of Procurement

Chapter 6:   Distribution ................................................................ 95
                    Introduction
                    Importance of the Distribution
                    function within the MM Context
                    Objectives of the Distribution Function
                    Elements of the Distribution Function
                    Channels of Physical Distribution
                    Transportation
                    Incoterms
                    Packaging
                    The Distribution Function in Organizational Structure

## Part 3: Technology Applications in Material Management

Chapter 7:   Technologies in Warehousing ..................................... 127
                    Introduction
                    The Importance of New Technologies in Warehousing
                    Warehouse Management System (WMS)
                    Applications of Technology in Warehousing
                    RFID Technology
                    Barcodes

                    Voice Recognition Technology  
                    Exchange Data Interchange EDI  
                    Enterprise Resource Planning -ERP  
                    Internet of Things - IoT  

Chapter 8:   Technology Applications in Procurement ................... 176  
                    Introduction  
                    e-Procurement  
                    Mobile Procurement - m-Procurement  
                    Blockchain Technology in Procurement  
                    Big Data in Procurement  
                    Negotiation via Videoconferencing  

Chapter 9:   Technology Applications in Distribution ....................228  
                    Introduction  
                    Control Tower Solution  
                    Intelligent Transportation System - ITS  
                    Global Positioning System - GPS  
                    Geographical Information System - GIS  
                    Distribution Requirements Planning (DRP)  

## Part 4: Technology Applications in Bahrain

Chapter 10:  Technologies in Material Management ......................260  
                    Introduction  
                    Technology in Bahrain  
                    Realising the Importance of Supply  
                    Chain by the Government  
                    Companies and Organizations in Bahrain (Private Sectors)  
                    Other Technologies used by the Private Sector in Bahrain  

Appendix ................................................................................283  
Index ......................................................................................287

# LIST OF FIGURES

Figure 2-1 Materials Flow through Logistic ..................................... 19
Figure 2-2 Centrally Managed Logistics ........................................... 30
Figure 3-1 Stages in TALC ................................................................ 37
Figure 3-2 First version of TAM (Davis et al., 1989) ...................... 41
Figure 4-1 The basic layout for a Warehouse ................................. 56
Figure 4-2 Cantilever Racking System ............................................. 57
Figure 4-3 Selective Racking System ................................................ 58
Figure 4-4 Push Back Racking System ............................................. 58
Figure 4-5 Drive-in Racking System ................................................. 59
Figure 4-6 Pallet Flow Racking System ........................................... 59
Figure 4-7 Carton Flow Racking System ......................................... 60
Figure 5-1 Common Organisational Structure of Procurement ....... 87
Figure 6-1 The Distribution function as part of Sales .................... 122
Figure 6-2 The Distribution function as part of Logistics .............. 123

# INSPIRATIONAL WORDS

*"The rapid technological advances being witnessed across the world, and thus businesses in Bahrain must benefit from Technology into finding the best solutions in different business sectors".*

Prince Salman bin Hamad Al Khalifa, Crown Prince, Deputy Supreme Commander and First Deputy Prime Minister of the Kingdom of Bahrain.

# PREFACE

a) **Subject**

This is a textbook about the key three pillars of Material Management, Warehousing, Procurement, and Distribution, and technology trends that being offered to improve these three functions. The book describes all activities and duties of each pillar in detail in order to let readers be familiar with the basic functions of each of them. It details the way that materials / stock items are acquired from suppliers, through the warehouse within an organisation, and then out to customers, which is an ultimate goal of being part of a supply chain, considering the chores and responsibilities of the three functions minutely. The book considers the Logistics as an essential base to control all three pillars and aspects within Material Management, and managers within three functions have to make the movement of materials as efficient and effective as possible. After explaining the basic responsibilities of (Warehousing, Procurement, and Distribution), the book, then moved to focus on the technology applications in each of them, bringing up all new technologies and innovations that have been developed to increase productivity in the three functions and minimise costs and errors. The author concentrated on the results that directly affect customer service, costs, productivity, quality of work, and achieving strategic objectives by implementing such technologies in Material management.

To emphasise the broader role of new technologies with Material Management, the book reviewed every single technology in each function and listed main advantages and disadvantages that companies need to take into consideration before implementing it. As Material Management is considered a fast-moving field based on rapid growing in technologies being offered, the book covered the technologies that currently being offered to transform the three pillars in Material Management, and continue to carve out their role in the global logistics industry. Importantly, all technologies covered in this book have already shown their usability and many companies in the worldwide have touched their greatest benefits.

This book gives an up-to date view of technology trends in Material Management and logistics, emphasising current trends and developments. As aforementioned, the book covers the main roles and duties of three functions within Material Management, along with the important technologies being offered. The following are the main topic presented in the book:

- Basic concepts and definitions of Management, Material Management, Technology, Inventory aspects, and Organisation.
- Details of main roles and duties of Warehousing, Procurement, and Distribution within organisations.
- Technology applications in warehousing such as Robotic systems, IoT, ERP, and RFID.
- Technology applications in Procurement such as e-Procurement, e-Commerce, e-Business, M-procurement, Blockchain, Big Data, and Video-conference.
- Technology applications in Distribution such the Control Tower, ITS, GPS, GIS, and DRP.
- Current technology applications in Material Management used by both private and public sectors in the Kingdom of Bahrain.

b) **Approach of the Book**

The book gives an introduction to logistics, Material Management, and technology trends that have the biggest impact on their operations and processes. It can be used by anyone who is meeting the subject for the first time. You might be a student taking a course in supply chain, logistics, Material Management, or Technology courses. The book is also useful for business studies, or to learn more about the main area of management. As aforementioned, the book gives a broad description of Material Management, covering all the main concepts and its technology trends, and it discusses the topics in enough depth to provide material for a complete course and enough to gain knowledge in book topics, but it does not get bogged down in too much detail. By concentrating on key issues, we have kept the text to a reasonable length. The book has a number of features. It:

- Is an introductory text and assumes no previous knowledge of new technology trends in Material Management.
- Can be used by many types of student, or people interested in book topics.
- Can be used by practitioners who need to be familiar with new technology trends in Material Management.
- Gives ideas about the current technologies used by private and public sectors in Bahrain.
- Can be an important material to help the kingdom of Bahrain to achieve its "Vision 2030".
- Takes a broad view, covering the main types of technologies in Material Management.
- Describes a lot of material, concentrating on topics that you will meet in practice
- Develops the contents in a logical order.
- Is practical, presenting ideas in a straightforward way, avoiding abstract discussions
- Demonstrates principles by examples and case studies drawn from international organisations and academic papers.
- Is written clearly, presenting ideas in an informative and easy style.

# PART ONE

Principles and Definitions

# CHAPTER 1

# Business Principles

**Management and Technology**

**1) Introduction**

Management is a concept plays an essential part of any organisation to set informal or formal flexibility policies and procedures to work in practice. Many scholars and authors have written about Management but the father of management is Peter F. Drucker, who died at 95 in 2005. He could revert as the father of modern management for numerous books and articles he published throughout his academic life. One of his distinguished approach in the field of modern management that he could stress innovations and smart technologies for dealing with a new challenging in the new business. Frederick Winslow Taylor, born March 20, 1856, Philadelphia, Pennsylvania, U.S., and died March 21, 1915, Philadelphia, is an American inventor and engineer who is known as the father of scientific management, is also known as a scientist of theory of management, and his theory is called Taylorism or the Taylor system.

In the course of technology history, there are different inventors who are known as fathers of Technology, who have helped shape IT with their inventions. The most inventors in recent years who Invented and

Innovated in Technology are: Andy Rubin (Smartphone OS Pioneer), Martin Casado (Networking Dynamo), Ivan Sutherland (Graphical Godfather), Alon Cohen (VoIP Visionary), John McCarthy (AI Architect), and J.C.R. Licklider (Cloud Prophet).

a) **Definition of Management**

The most elaborate definitions which represent the wide spectrum of management concepts are listed below:

- 'Management is an art of knowing what to do, when to do and see that it is done in the best and cheapest way.' (Taylor 2008).
- 'Management is a science that implies the idea / ideas are 5 main functions of designing, ordering, organizing, controlling and mengoordinasasi.' (Barnard 1938).
- 'Strategic Management is what matters for the effectiveness of the organization, the external point of view, which stresses the relevance of the objectives against the environment, in terms of internal stresses, the balanced communication between members of the organization and a willingness to contribute towards actions and the achievement of common objectives.' (Barnard 1938).
- 'Managers are the people to whom this management task is assigned, and it is generally thought that they achieve the desired goals through the key functions of planning and budgeting, organizing and staffing, problem solving and controlling. Leaders on the other hand, set a direction, align people, motivate and inspire' (Kotter 2001).
- 'knowledge Management is an important element in the management concept, it refers to the leveraging of the organizations' collective wisdom (know-how) by creating systems and processes to support and facilitate the identification, capture, dissemination and use of the organization's knowledge to meet its business objectives.' (Lastres 2011).

As a common definition as revealed in the definitions that Management means a specific process of planning, organizing, directing and controlling the efforts of the people along with the required resources that are engaged in business activities in order to attain predetermined objectives of organizations.

b) **Definition of Technology**

Technology was defined as the spoken word of manual craft or cunning skill in the ancient time. The earliest use of the word technology in the United States was found in a Harvard University course on the "application of the Sciences to the useful Arts" in 1816. The 1832 Encyclopedia Americana defined technology as principles, processes, and nomenclatures. The advancements in technology or modern technology have brought many changes to the life styles of people. It has pervaded every aspect of human life, whether it is health, education, economic, governance, entertainment, etc. Thus no matter what the field is, technology must have brought some positive change to work in a way to increase productivity. Today, every nation strives to get the latest technology for the benefit of its citizens. Technological progress is vital in the fields of business, education, health care, and other aspects of community. Technology is also seen as an enabler or a vehicle to disseminate knowledge.

The following quotes represent the main definition of technology thinking, which are relevant to the technology aspects related to the book topics:

- ➢ 'A system created by humans that uses knowledge and organization to produce objects and techniques for the attainment of specific goals.' (Volti 2009).
- ➢ 'Technology as a word-root is traditionally understood to refer to "art" or "skill" (Skrbina 2015).

- ➤ 'Technology consists of two primary components: 1) a physical component which comprises of items such as products, tooling, equipments, blueprints, techniques, and processes; and 2) the informational component which consists of know-how in management, marketing, production, quality control, reliability, skilled labor and functional areas.' (Kumar 1999).
- ➤ 'Technology is always connected with obtaining certain result, resolving certain problems, completing certain tasks using particular skills, employing knowledge and exploiting assets.' (Lan 1996).
- ➤ 'Technology is the firm's 'intangible assets' or 'firm-specific' which forms the basis of a firm's competitiveness and will generally release under special condition.' (Dunning 1994).
- ➤ 'Technology as the theoretical and practical knowledge, skills, and artifacts that can be used to develop products and services as well as their production and delivery systems. Technology is also embodied in people, materials, cognitive and physical processes, facilities, machines and tools.' (Burgelman 1996).
- ➤ 'Technology transfer as transmission of know-how (knowledge) which enable the recipient enterprise to manufacture a particular product or provide a specific service.' (Baronson 1970).

## 2) Material Stock /Inventory

Inventory (American English) or stock (British English) is defined by prominent Institutes of Logistics as all the goods and materials held by an organization for sale or use, or a list of items held in stock. There is different between stock materials and non-stock materials in terms of their recording and controlling. Non-stock Materials are not held in stock and properly they are consumed immediately, while doing the goods receipts.

- ➤ 'Stock items are goods that held in a sufficient level at minimise waste and to optimise the overall cost of the holding stock to meet the needs of consumers.' **CIPS**[1].
- ➤ 'Stock is used to support production (raw materials and work-in-process items), supporting activities (maintenance, repair, and operating supplies), and customer service (finished goods and spare parts.' **APICS**[2].
- ➤ 'The purpose of inventory theory is to determine the rules that management can use to minimize the costs associated with maintaining inventory and meeting customer demand.' **CILT**[3].
- ➤ 'Inventory is a usable but an idle resource having some economic value.' **CII**[4].
- ➤ 'Inventory speaks to a huge cost to the humanitarian supply chain. One of the essential differences is in the multifaceted nature of each kind of inventory systems.' **IISCM**[5].
- ➤ 'Inventory is goods that controlled and planned by an organisation for consuming into, through, and out of, an organization.' **CPIM**[6].

## The Aspects of Materials Managemet

## 3) The Stock Control

The stock control is a vital technique within the aspect of materials management. This technique is about how to control inventories in a warehouse as an essential inventory management to ensure the effectiveness of the operation of an organisation. The stock control technique and process ensures an appropriate stock level is maintained by an organisation in order to be able to meet customer demand

---

[1] Chartered Institute of Procurement and Supply
[2] American Production and Inventory Control Society
[3] Chartered Institute of Logistics and Transport
[4] Confederation of Indian Industry - Institute of Logistics
[5] Indian Institute of Supply Chain Management
[6] Certified in Production and Inventory Management

without delay while keeping the costs associated with holding stock to a minimum. The stock control aims to achieve the following advantages for an organisation:

i) It ensures an efficient inventories held in an organisation and only the required quantities are stored to meet the business purpose.
ii) It ensures lower storage costs.
iii) It ensures lower management costs.
iv) It ensures up to date stock records and avoid wasting stock.
v) It ensures accurate reviews of stock for placing an order to return stocks to a predetermined level.
vi) It ensures the right amount of inventory in the right place.
vii) It ensures stock are organised with easily understandable SKUs.

## 4) Stock Planning

The stock planning is an essential process in materials management for determining the optimal quantity and timing of inventory for the purpose of aligning it with sales and production capacity. Inventory planning has a direct impact on a company's cash flow and profit margins since it affects on turnover of materials. The stock planning process is always aligned with the organisation' strategy, and It also impacts the whole supply chain activities as it represents the value an organisation provides to their market. Without the stock planning or lack of correct planning could lead to:

i) Decreases productivity and outcomes in warehousing.
ii) Increase storage costs.
iii) Delay in shipping of materials, and inaccurate shipments for the same materials and purchase orders.
iv) Could cause loss customers if materials are not available for sales, and unacceptable levels of backorders.
v) Cause a high risk of suffering considerable damage to an organisation's reputation in the market.

vi) Cause a lower return on investment from organisation's warehouse and inventory.
vii) Higher costs of major projects due to late deliveries of raw materials and finished goods.

## 5) Stock Records

The stock records process is a manual or computer-based record of the quantity and kind of inventory in the stock/book, and it's used to ensure accuracy of stock in the warehouse. The records include history of the recent transactions take place in each stock/inventory item on day to day activities. The stock process ensures effective inventory use in a warehouse, and accurately reflect the quantity of materials available for use in Production and Sales. The stock records, including type, date range, format, volume, storage location, and applicable records series information. Maintaining accurate records for stock is essential to running an efficient and effective organisation, and inaccurate or invalid information, management might be unable to make decisions about the company and can result in significant business problems including:

i) Inaccurate inventory records can result in lost business for Production (raw materials not available at time of operations) and Sales (finished good not available at the time of sale).
ii) Disrupted operations since machines and resources become idle since the required materials are not available for production.
iii) Poor customer service (late deliveries to customers).
iv) Lower productivity (additional setups to complete a job).
v) Poor material planning (the inventory records are critical in determining MRP quantities).
vi) Excessive expediting (trying to obtain necessary items in less than normal lead time).
vii) Increase ordering costs (items could be sold more than their normal price by suppliers).

## 6) Materials Handling

Material handling system is a concept used to describe the movement, protection, storage and control of materials and products throughout manufacturing, warehousing, distribution, consumption and disposal. As a process, material handling incorporates a wide range of manual, semi-automated and automated equipment and systems that support logistics and make the supply chain work. Material handling system is an essential in manufacturing and logistics and plays as part of every warehousing worker's job. It also plays an important role to improve customer service, reduce inventory, shorten delivery time, and lower overall handling costs in manufacturing, distribution and transportation.

Material handling system involves diverse operations such as hoisting tons of steel with a crane; driving a truck loaded with concrete blocks; carrying bags or materials manually; and stacking palletised bricks or other materials such as drums, barrels, kegs, and lumber. The process is essential in logistics to provide a continuous flow of parts and assemblies through the workplace and ensure that materials are available when needed for production, and to ensure loading items for dispatching. The advantages of material handling systems in logistics and warehousing in particular are:

 i) Increase productivity.
 ii) Save time and cost.
 iii) Reduce materials inventory.
 iv) Provide better working condition.
 v) Make improvement in quality and reduce damage to materials.
 vi) Flexibility in terms of using manpower.
 vii) Improve delivery services.

## The Warehouse

## 7) The Warehouse Manager

Warehousing's role is a critical part of Materials Management. The role involved from maintaining long-term storage of stock (raw materials, WIP, finished goods) to supporting the economies of purchasing, production, and transportation including light manufacturing and facilitating time-based supply chain strategies. The warehousing role contributes to the overall processes of a supply chain as well as the total cost, and it takes a responsibility to trade-offs between warehousing costs and services to other critical functions involved within the supply chain. The main responsibilities of the warehouse manager are:

i) Contributes to reduced storage and resource costs.
ii) Improve services to internal users (other functions) and external users (customers).
iii) Make the operations and processes more flexible, efficient, and responsiveness to add more value to the organisation and supply chain as a whole.
iv) Ensure storing product in the right location with the right records to fulfill both internal and external users demand, and protect against uncertainties in demand and lead-time.
v) Ensure an effective communication with other functions within the organisation to enable an inseparable part of supply management and make higher performance and higher productivity.
vi) Ensure basic management functions are fulfilled in terms of Planning, Organizing, Leading and Controlling.
vii) Maintain and improve the material handling system in the warehouse through enhancing its productivity, utilisation, and safety aspects.
viii) Improve stores facilities for ensuring product accuracy, quantity, timing of shipment and delivery, accuracy of documentation, and overall product condition.

## The Procurement / Purchasing

## 8) The Procurement Manager

The role of procurement or purchasing in Material Management is very critical as it has a direct impact on two of the most important factors that drive a company's bottom line: cost and sales. The procurement function is responsible to obtain materials, lease or other legal means, parts, supplies, equipment, and services required by an undertaking for use in an organisation. The main critical aspect of the role is to ensure obtaining the above things at the right time, right quantity, right quality, right time, and right price. In addition to the mentioned responsibilities, the procurement manager has other duties as follows:

i) Develop supplier relationship.
ii) Garner cost savings for the organizations without trading off quality and quality.
iii) Conduct negotiation with suppliers and vendors on price and other aspects related to contracts and purchase orders.
iv) Expedite receipt and payment with suppliers.
v) Handle the management duties of the various resources in his/her department and coordinate them properly.
vi) Coordinate with other functions, especially with the warehouse manager over the inventory stock of the company, in order to ensure all the products are available and ready in a timely manner.
vii) Involve in paperworks works such as running reports, documents, and analysis.

## The Distribution Management

## 9) The Distribution Manager

In some companies the distribution and warehousing roles have the same activities, but the difference in their purpose. However, the key role of the distribution manager is to ensure goods are distributed to the customers as the orders are received. The following are the main activities of the distribution manager in most companies:

i) Set distribution goals and plan and manage distribution operations to achieve the set goals.
ii) Identify resources, perform workload assignments and provide assistances when required.
iii) Conduct trainings to team members as needed.
iv) Monitor team performances and provide feedback for improvements.
v) Oversee daily routes and improve route management to ensure timely deliveries.
vi) Follow and enforce company policies and procedures.
vii) To address customer queries and ensure customer satisfaction.
viii) Develop distribution budgets and manage expenses within the budgets.
ix) Work with warehouse and transportation staffs to ensure timely and accurate deliveries.
x) Perform cost negotiations with vendors and provide appropriate solutions for bulk distributions.
xi) Ensure that company transport vehicles are maintained in good working conditions.
xii) Develop cost reduction initiatives while maintaining productivity and quality.
xiii) Plan and schedule deliveries to meet customer needs.
xiv) Assist in inventory management activities including receiving, storing, rotating and handling supplies in distribution center.

## 10) Definition of Organisation

Organisation is defined as a social unit of people that is structured and managed to meet a need or to pursue collective goals. All organisations have a management structure that determines relationships between the different activities and the members, and subdivides and assigns roles, responsibilities, and authority to carry out different tasks. Organisations are open systems which affect and get affected by their environment. There are various goals of organisational management, but the main goal is to use the various levels of company leadership in the leadership hierarchy to set goals, monitor results and build a stronger company. Strategies might involve employee training, promotional strategies, operations efficiency or any other aspect of the company.

## Main Features and Functions of Organisation

a) **Planning**

'The planning function is typically where the direction of the organization is established through a variety of activities including the development of goals' (Leung 2004).

i) Prepare an effective business plan for all functions. It is essential to decide on the future course of action to avoid confusions later on.
ii) Plan the business strategy for the organisation.
iii) Develop the vision and mission of the organisation.

b) **Organising**

'The organising function of management is comprised of numerous activities directly or indirectly related to the allocation of resources in ways that support the achievement of goals and plans that were developed in the planning function' (Leung 2004).

i) Organise appropriate and judicious resources to achieve the best out of the employees.
ii) Organise and design of individual jobs within the organisation.
iii) Prepare a monthly budget for smooth cash flow.
iv) Organise the business plan and procedures for all functions.

c) **Staffing**

'It is the function of management provides a workforce that can offer a varied resource of creativity and flexibility to organisation' (Sawhney 2013).

i) Recruit the right manpower and talent for the organisation.
ii) Monitor and evaluate the performance of functions and staff and adjust activities depending on results.
iii) Prepare training, compensation and appraisal for all staff based upon the information given by their respective department.

d) **Leading**

'In the context of leadership, trust plays a valuable role in the job attitudes displayed by employees in the work environment. Recent research indicates different aspects of trust are related to higher levels of many of these job attitudes, to include perceived organisational support and affective organisational commitment' (Ferres 2004).

i) To lead all functions through managers or superiors in each function by setting clear targets for each one.
ii) To lead the team members work in unison towards a common objective and decide the right action in a particular situation.

e) **Controlling**

'It is about managerial efforts directed toward monitoring both organisational and employee performance and progress toward goals' (Costa 2007).

i) To ensure a targeted element of performance remains within acceptable limits.

ii) The organisation establish standards and then measure performance against the set standards, and then takes correcting deviations from standards and plans.
iii) Control involves managing the organisation's process, including the financial control such as debt, cash flow and receivables/payables, and budgetary control which is considered the most important control of all.

f) **Motivating**

'It is a part of leadership which is a multi-dimensional process, including motivating employees' (Howell 2006).

i) Motivate staff towards behaving business goals.
ii) Motivate the employees to achieve organisational tasks.
iii) Appreciate the employees for their good work or lucrative incentive schemes go a long way in motivating the employees and make them work for a longer span of time.
iv) Motivate the employees through pay bonus, rewards for exceptional achievement, and promotion among others.

## 11) Definition of Information Technology (IT)

Information Technology which is pronounced "I.T." refers to stuff related to computing technology, such as networking, hardware, software, the Internet, or the people that work with these technologies. In current competitive markets, many organisations, especially in developed countries obtain latest technologies to keep their business distinguished from their competitors. For that reason, IT departments in an organisation play an important role for managing the computers, networks, and other technical areas of their businesses, and HRD always tries to recruit IT specialists who have the ability and sufficient skills to handle such critical tasks. Information Technology (IT) in current business is evident through its applications to strengthening the elements of the competitive advantage in terms of business expansion, cost reduction, time savings, along with high flexibility to develop the performance to respond to customer demands. Information system (IS) is similar to Information Technology (IT) in many ways, but at the same time they are different in some aspects, which depends the way to utilise each one. The field of information systems links the computer functions to businesses, through developing a system to meet all business purposes. Although information systems are heavily reliant on computers and other technology-based tools, the term predates computers and can include non-technological systems. Examples of information system in Material Management are Enterprise Resource Planning (ERP), Product Data Management (PDM), Customer Relationship Management (CRM), Supply Chain Management (SCM), and others.

In Material Management, IT has become an important factor to face the new challenges in warehousing, distribution, and procurement according, to many studies. By adopting new technologies in Material Management, new technologies have made easier to control and handle operations and processes in material management, and found that modern technologies have helped to meet business objectives for organisations. There are five main components for any Information Technology:

a) **Hardware:** The term refers to the physical components that make up a computer or electronic system and everything else involved that is physically tangible. The computer hardware includes the monitor, hard drive, memory and the CPU. It also includes the input methods, the output means, various means of storage, and means of communications.

b) **Software:** Is a set of a generic term used to describe computer programs. It is developed through a set of instructions or programs instructing a computer to do specific tasks. It includes two types of programs: programs for operating the hard entity and the software applications which the final user deals with. Software is divided into three categories:

i) **System software** serves as a base for application software. System software includes device drivers, operating systems (OSs), compilers, disk formatters, text editors and utilities helping the computer to operate more efficiently. It is also responsible for managing hardware components and providing basic non-task-specific functions.
ii) **Programming software** is a set of tools to aid developers in writing programs. The various tools available are compilers, linkers, debuggers, interpreters and text editors.
iii) **Application software** is intended to perform certain tasks. Examples of application software, include office suites, gaming applications, database systems and educational software. Application software can be a single program or a collection of small programs. This type of software is what consumers most typically think of as "software".

c) **Communication Networks:** It is a critical part of information technology, which facilitates interpersonal communications, allowing users to communicate efficiently and easily via various means: email, instant messaging, online chat, telephone, video telephone calls, and video conferencing. A network allows sharing of network and computing resources. Communication network's role is to exchange data, information, knowledge, and software among people through certain means of information technology and within different networks. Each of these networks work on a limited scale and the other covers a wide geographical area. Data and information exchange can be limited to a certain scale between people and organisations or can be made accessible to everyone.

d) **Database:** It is a collection of related data that is stored in a manner enabling information to be retrieved as needed; in a relational database, a collection of related tables. Database management system (DBMS) is a type of software program used to create, maintain, and access databases. A database system is a repository for the data, themes, and organised and interconnected files with each other. It describes the current and previous operations of the organisation, which can be referred to quickly by computers associated with various programs. The database can be added, modified, and modernised to keep pace with constantly evolving variables. It enables managers to make strategic decisions in accordance with the right basis, and it also enables users to do their jobs more efficiently.

e) **People:** People are the most important structural element in the information technology system. The people that are needed to run the system and the procedures they follow so that the knowledge in the huge databases and data warehouses can be turned into learning that can interpret what has happened in the past and guide future action. People can be divided into two categories. The first, which is the dominant are called the end users, who deals with the application programs such as beneficiaries of it and its applications without giving precise details of the programming operations. The second class is the specialists in the field of computer who design computers and installs different programs on them, whether they are application software or system software.

# CHAPTER 2

# Logistics

## 1) Introduction

The concept of Material Management is known as controlling the flow of materials, from supplier through production to consumer/customer, and for that reason companies like to use this term for emphasis the key functions that must be conducted to minimise total costs in this area and provide a better level of customer service, along with coordinating activities and processes among all stakeholders for better planning and controlling materials flow. To this end, Logistics is considered the main area to control material flow, operations, and processes in order to achieve an effective materials management system that could integrate the entire material and supply chain work processes.

## 2) Logistics within the Supply Chain Management

Logistics is one of the critical important part of supply chain management. Supply chain management is the term used to define the control and organised flow of material, information, people, and finances as they budge in a process from supplier to manufacturer to the end customers (wholesaler to retailer, or individuals). In other words, it is a process of upstream and downstream relationships between suppliers and customers to meet customer demands. There are a number of stages are involved in a supply chain, typically including suppliers, manufacturers, warehouses, distribution centres, retailers, and customers. The material flows move through the Logistics as the key part of supply chain from suppliers to customers, while the information flows of orders and demands are in an opposite direction *(see figure 2.1)*. The supply chain is normally divided into three flows such as i) the product flow ii) the information flow iii) the finances flow. The finance flow could be flown in both directions, but the product flow starts from the supplier to the customer and the information flow related to the fill up the demand, location providing services and feedbacks starts from the end customer through customer demand orders.

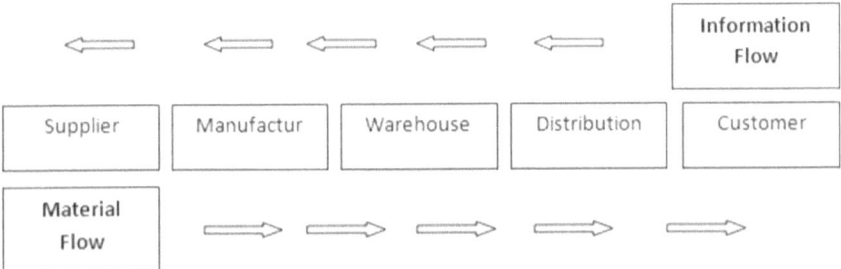

Figure 2-1 Materials Flow through Logistic

The supply chain plays an important role in today's highly competitive and complex marketplace to face a challenge through its effectiveness and efficiency, and most companies in today's business try to have more advantage than its competitors by applying new technologies throughout all chains to make jobs easier to track flow of all the activities in supply chain management. Another important role of supply chain includes the coordination and collaboration between all partners such as suppliers, manufacturers, warehouses, third-party service providers (Distribution companies), and customers, and that happens through an integrated system.

There are some key elements must be established to make the supply chain achieve organisational goals:

i) **Service Level Management:** Ensure all service management processes and operations in the supply chain are meeting the service level targets. Prioritising goals are important to set and review regularly.

ii) **Order and Demand Management:** Ensure orders and demand management are properly arranged and controlled, and proper systems are used to track them with both suppliers and customers. Planning and forecasting are the most important methods that can be applied to ensure proper order and inventory controls.

iii) **Production Management:** Ensure any activity related to production is planned, organised, and controlled properly. Managing production activities means managing and coordinating the 6M's i.e. men, money, machines, materials, methods and markets to satisfy the needs of the organisation.

iv) **Supply Management**: It is the responsibility of the purchasing function to ensure all activities related to making a relation with suppliers. It must be managed properly in terms of identifying, acquiring and managing the resources and suppliers that are essential to the operations of an organisation. One of the main objectives that must be considered in any relation with suppliers are the cost of purchases, the efficient allocation of resources, risk management, and the effective gathering of information to be used in strategic business decisions. The concept is also known as "procurment", which includes procurement planning, supplier performance, and supplier relationship.

v) **Distribution Management**: This is one important activity that must be acted properly to ensure proper movement of products from the supplier or manufacturer to the point of sale. The management includes numerous activities and processes such as packaging, inventory, warehousing, supply chain and logistics. With new technologies, the distribution management has become more savvy through easier gathering and sharing of relevant information between partners in identifying key opportunities for growth and competitiveness in the current market. The activity includes commercial distribution (commonly known as the distribution centre for sales) and physical distribution, better known as logistics. There are different distribution management methods used to perform the activity to ensure operations within the supply chain are properly performed for customer service, shipping, warehousing, inventory control, private trucking-fleet operations, packaging, receiving, materials handling, and plant, warehouse, store location planning, and the integration of information. The manager's role in this respect is to ensure the distribution goal is achieved in the delivery of raw materials, parts, and finished products to the right place and time, in the proper condition with the right transportation method.

vi) **Integrated SCM planning and execution:** All processes within the supply chain must be integrated through interdepartmental interaction and interdepartmental collaboration to bring all departments involved together into a cohesive organisation. The planning and execution must be set strategically, tactically and operationally to involve suppliers, manufacturers and customers. In other words, it must cover both internal and external integration:

*"Internally, logistics must work closely with production and marketing, to plan, coordinate, and integrate the cross-functional activities to create value for customers. Externally, logistics can serve in a boundary-spanning capacity and interface with suppliers, carriers, and customers."*

(Chen, Mattioda and Daugherty 2007)

With new technologies and information systems, managers can optimise the integration and execution by having deterministic supply chain planning and processing.

## 3) Designation of Logistics

Some of the most popular definitions and designation of logistics cited by professional and prominent bodies are:

- ➢ 'Logistics is the flow of funds, goods and information between origin and consumption. Logistics involves information, material handling, production, packaging, inventory, transportation, warehousing and often security.' *CIPS (2018)*.
- ➢ 'Logistics is the process of designing, managing and improving such supply chains, which might include purchasing, manufacturing, storage and, of course, transport.' *CILT (2012)*.
- ➢ 'Logistics as that part of supply chain management that plans, implements, and controls the efficient, effective flow and storage of goods, services, and related information from the point of origin to the point of consumption in order to meet customers' requirements.' [7]*CSCMP (2018)*.
- ➢ 'Logistics is the process of planning, implementing and controlling the efficient, cost-effective flow and storage of raw materials, in-process inventory, finished goods and related information from point of origin to point of consumption for the purpose of conforming to customer requirements.' [8]*ISM (2018)*.

## 4) Task of Logistics

In Modern logistics the tasks and roles have been increasingly applied in interdisciplinary science. The knowledge and information are the main elements of any task carried out in logistics as well as in other disciplines, in a way to deal with the whole processes such as storing, processing, transferring and providing data and information. The task of logistics is to develop and organise optimal processes, structures, systems and networks for the operational logistics. The main tasks of logistic

---

[7] The Council of Supply Chain Management Professionals
[8] Institute for Supply Management

management are to execute the orders and fulfill the requirements of consumers and companies at the lowest costs with adequate quality. The logistic activities represent one of the highest costs of doing business, and in some studies the costs represent between 5% to 35% of sales depending on the type of business. However, new technologies could reduce logistic costs through switching to new methods to improve supply chain processes and in turn save a business money. There are several methods that companies use to control or reduce logistic costs such as Just In Time, Third party logistics (3PL), automated warehousing, intermediate bulk containers (IBCs), integrated material handling systems, and/or through strengthening relationships with suppliers through the real time systems (e.g. B2B or B2C).

## 5) Integrated Logistics Platform

The Logistics platform is the new term used to define the relationship and coordination between the key functions within the supply chain. The platform is a specialised area with the infrastructure and services for coordination and collaboration the distribution activities for achieving added value services. The Logistic platform is literally defined as:

*"An area with which all activities relating to transport, logistics and the distribution of goods, both for national and international transit, are carried out by various operators. It is run by a single body, either public or private, and is equipped with all the public facilities to carry out the above mentioned operations"* (Europlatforms 2004).

The key logistics activities are coordinated and managed through one network to carry out the following:

i) Logistics operations.
ii) Physical structure, processes, activities.
iii) Information systems for design, operations and reporting.

The main goal of the logistics platform is to provide better services to customers with less costs, relying on new information systems through integrating information flowing between all stakeholders. The network is required all partners to have specific integrated systems to manage the intentions and interactions for an easier approach to each other, and new technologies play an important role in this respect. The logistic platform works through the interaction between logistics and other channels in the same organisation and with other players such as suppliers and distribution centers to a key determinant for the organisation's success, which means transport, logistics and the distribution of goods are the ultimate goals the logistics platform to increase efficiency, inventory reduction, increased sales, building relationships and improved customer service within the supply chain management.

## Types of the Logistics Platform

i) **Unimodal distribution area**: This type is related to a unimodal distribution centre as storage facilities such as Warehouse to manage product flows and associated stocks. This type of platform is largely aimed at the management of product flows and associated stocks. The unimodal infrastructures type is operated by one or several firms, and do not necessarily involve joint operations, and the main objective is to optimise operations of road transport.

ii) **Logistic areas**: This type involves more integrated operations (not only road transport) to include stock consolidation, local and re-directing activities. The platform is normally developed to undertake activities in traffic and freight division points for switching to different modes of transport, and include more than one mode of transport, enabling distribution rearrangement and cross-docking activities. A good example of this type is the method of activity conducted in air and/or maritime freight centres.

iii) **Multimodal platforms**: This type of platforms is built through connecting different modes of transport, and focusing on added value services through linking to ports, which is known as hubs. Such infrastructure is aimed to make the most of scale economies on international routes. Their nodal function includes national and international logistics and transport-related activities, and run by several operators. Due to the large volumes handled and their excellent locations, they enable the implementation of nearly all the different postponement strategies (geographic, manufacture and assembly).

## Advantage of the Integrated Logistics

The role of logistics includes activities such as transportation, warehousing, inventories, order processing, packaging, materials handling, forecasting and planning and purchasing, and all the functions must be integrated to add value through transportation, holding accurate inventory management, better utilisation of time and space, allowing for the necessary quantity of products to reach all points in the chain efficiently, cost-effectively and in the timely manner. The new logistic performance has a significant impact on the economy of a country from a micro-economic point of view on company profitability and performance, as the focus has been switched from cost reduction to value generation. With growing of Information Technology in recent years, the integration between functions inside the organisation and with other partners has become more efficient by sharing the same purpose and objectives through a holistic solution. The ERP system is a good example which is able to communicate all functions together to overcome traditional functional barriers to achieve the same goals. The main benefits of integrated logistics are:

a) **Enhancement in:**

　　i) Efficiency in terms of meeting deadlines, managing costs and delivering quality.

ii) Efficiency in highly organised storage and stock management, and fast and accurate picking and packing, and distribution of the right goods to the right place at the right time.
iii) Better supplier administration.
iv) Complementarities between companies.
v) Improved productivity.
vi) Improve service levels to customers.
vii) Increase competitiveness of the firm.
viii) Standardisation of procedures.
ix) Enhance communications internally and externally.
x) Improve visibility through an integrated ERP system.

b) **Cost saving benefits**:

i) Holding the optimised stock in a warehouse.
ii) Reduction in lead time for better customer services.
iii) Reduction in buffer stock.
iv) Reduction in resources and facilities.
v) Reduction in manpower and headcounts.
vi) Reduction in transportation and fuel costs.
vii) Reduction in freight costs.
viii) Avoiding high insurance costs for storage and stock.
ix) Avoiding delivery delays.

## 6) Logistics and Materials Management (MM)

Materials Management deals with the physical items that are needed for producing goods and services, and there are three types of materials considered significant to the key operations, especially in the manufacturing industry; the raw materials, the work in progress (WIP) inventory and the finished goods. Other types of materials are used as parts and tools for factory machines and miscellaneous works. An effective materials management within the logistics will result:

i) Better control of materials.
ii) Efficient management of storehouse and store-yard.
iii) Lower logistics costs.
iv) Efficient business communication.
v) Effective logistics management.
vi) Reduction in the overall cost of the materials.
vii) Efficient materials handling.
viii) Reduction in duplicate orders.
ix) Improvement in the quality of the materials.
x) Reduction in the inventory needed to be stored.
xi) Improved turnover of the materials.
xii) Improved economy in the purchase of the materials.
xiii) Improved relationship with the suppliers.
xiv) Better cash flow management.
xv) Improved teamwork and relationship with production departments.
xvi) Materials are available in stores when needed and in the quantities required.
xvii) Improvements in the labour productivity.
xviii) Improvements in overall plant availability for production.

## 7) Logistics and Distribution Management

Distribution is part of the logistics process through physical distributions of the raw materials, parts, and finished products throughout the logistic channels. The distribution activities take place through three flows (physical resources, information, and money), which are normally concomitant with each other. With new technologies, the distribution process has become simplified through bridging the gap between stakeholders in terms of communications and coordination using methods such as B2B or B2C. Therefore, more cost-effective, shorter use of time, more safety and higher level of executive capacity and many other business benefits could be reached by the logistics. An effective distribution management within the logistics will result:

i)   Decrease costs of distribution.
ii)  Efficient transportation mode choice.
iii) Efficiency in real time freight rates by making the best choice of shipments.
iv)  Increased customer service.
v)   Better logistics processes control.
vi)  More scalability and speed in arranging the distribution of physical resources.
vii) Increase business friendly with partners.
viii) Improvements in packaging.
ix)  Easier in route and vehicle scheduling.
x)   Easier in tracking and tracing orders.
xi)  Better mobility and flexibility in shipping goods.

## 8) Logistics and Procurement

Procurement or Purchasing is the key part of logistics as it has responsibility for managing a company's incoming material need for manufacturing or end users. The procurement's role is to order goods and services, obtain bids from third-party logistics providers (3PLs or 4PLs), create and negotiate of contracts with suppliers and service providers, and rate the vendors and suppliers. There is a strong collaborative system between procurement and logistics, which covers the key integrated supply chain processes for better business results. According to many studies, together, purchasing and logistics represent 70% of an organisation's costs and influence 80 percent of its working capital through inventory and accounts payable, which indicates the impact of procurement on the firm's overall performance. Breakthroughs in technological advances in logistics have made the purchasing processes easier. The procurement function is building the relationship with suppliers and other partners in logistics through using the latest information systems such as ERP, B2B, and P2P systems. An effective procurement within the logistics will result:

i) Achieve high levels of effectiveness when making inquiries, ordering and buying components.
ii) Reduction in storage area costs.
iii) Minimisation in tied capital.
iv) Building long-standing business relations with suppliers based on win-win.
v) Increase levels of functional and financial performance.
vi) Improve efficiency in the whole logistics process.
vii) Reduction in complexity.
viii) Lower operating expenses and cost of goods sold and inventory.
ix) Well aligned to the business unit's strategy and activities.
x) Avoid stock out.
xi) Improving in the lead time.
xii) Achieve better deals with suppliers.

## 9) Logistics in Organisational Structures

The corporate organisational structure is usually designed to create the best possible environment to achieve business objectives. Therefore, each company creates and tailors the organisational structure according to its core objectives to support the implementation of corporate strategy, and achieve stability and transparency in the functioning and fulfilling tasks of the organisation. However, in the current competitive market, the organisational structure is considered to be more flexible and adaptable to ensure more efficiency to cope with the rapid changes in market conditions. For that reason, the current trend and choice is to implement the lean organisational structure to face new challenges with respect to meeting customer demands. The logistics one of the core functions that needs to be tailored in the structure to ensure all activities are linked together in a functionally oriented matrix arrangement in order to make information flows through departments effortlessly. In other words, the logistics activities are increasingly perceived in the organisational structures to ensure well managed of the flow of materials throughout the whole company, starting from the point of receiving the customer orders, enter requests into the production plan, order components based on the production plan, they manage the input material stores, final production warehouses, and finally dispatching the goods to the customers. *Figure 2.2* shows the location of the logistics function in the organisational structure, which is normally applied in the manufacturing industry to ensure centrally managed all activities and information.

Figure 2-2 Centrally Managed Logistics

# CHAPTER 3
# Technology

## 1) Introduction

New technology has made the life more convenient in all aspects, and for that reason companies in the current business can't progress or sustain without stretching out on the new technologies. New technology has constantly thrown up new and innovative ways of altering the work environment, providing the opportunities for all types of business to become more efficient. Today, many leading organisations in business and government are considering new technologies as a business strategy. However, the key to success of using any type of technology in business processes depends on the implementation and understanding the way to drive true change to the business.

Many businesses have transferred from the traditional methods to the automated methods by using and implementing new technologies, including logistics as the engine that drives businesses to create greater visibility within the supply chain, gaining more control over the inventory, reducing operating costs, and outpacing the competition. Material Management in the logistics industry is considered a massive market, and forms the largest industry worldwide, as cited in recent studies, and its effect in a wide range in all types of business. Furthermore, the processes within logistics in today's business are considered more links either internally or externally, which could make more convoluted in facing new challenges if companies ignore new technologies, and then difficult to eliminate unnecessary links for keeping up with the market. Also, with the new technology bandwagon, companies are able to create predictable, consistency and visibility, which is the way to build faster communications and more efficiently with other stakeholders, because there is no excuse to convince customers if not being served as they are embracing connected 24/7. For that reason, many companies around the world have realised that an effective logistics process is not possible to achieve without considering the new technology methods, as innovation has become more pervasive to sustain the business. For all the mentioned reasons, new terms have occurred in SCM in recent years such as supply chain information systems (SCIS) to indicate a tremendous impact of information technology on the material management and logistics operations. According to some studies, new technologies have the capacity to impact organisational structure, firm strategy, communication exchange, operational procedures, inventory process, and supplier relationships, as the logistics management is recognised as an important area for information technology innovation and investment.

## 2) Initial Stage of Technology Adoption

Understanding concepts of technology application is crucial in getting a clear understanding of the nature of technology that must be implemented in logistics and business as whole. However, before thinking of adopting any type of technology, it is important for management to have the basic understanding of what material management and logistics entail as the first step to adopt the successful technology, and hence achieving the success of the business. The nature of the material management and logistics environment is a complex system of activities, involving sourcing, manufacturing, distribution, dispatching, consumption, and disposal, and everyone has a specific role to play. Moreover, the processes associated with logistics involving people, organisations, technologies, and control of materials and associated information across the life cycle of any ordering process, for all that reasons adopting and implementing any type of technology require particular skills and knowledge. As a first step, it is important to identify the goals of having new technologies in logistics by consulting the concerned people and functions, along with the development team. The phase is considered the conception stage to understand carefully both feasibility and value of the potential technology to the organisation. Having fully convinced with all details in step one, then proceeding with the next phase to have a full idea of how long it will take to make, how much it will cost, and it's worth pursuing, then the final decision is taken by top management to go ahead with the project and hence move onto the project life cycle undertaken by the project team.

## 3) Technology Development Life Cycle (TDLC)

The technology life cycle is a crucial analysis should be conducted to be familiar with the visibility of the technology being considered to adapt. Such the stage is very important before the real work and development of the system begin, as it predicates the journey the potential technology takes. The analysis indicates the system from its exciting birth and growth; moving to its inevitable decline and eventual death. Also, the assessment is able to predict when an organisation is able to recover the investment put into its development, and when to plan for a new project. One of the methods which is normally used throughout the life cycle process is the TDLC, which known as scrum to help structure more complex development projects. The TDLC is an acronym stands for Technology Development Life Cycle, is normally conducted as a process to plan, define, test and implement any new technology being considered to use or implement, in order to ensure success of the technology by driving a defined set of activities through each phase of the model. The TDLC highlights different phases of the development process to let users understand what activities are involved within a given step. The main phases of the TDLC are:

a) **Planning:** In this phase the development team discusses with the stakeholders the requirements of the technology need to be developed to achieve the potential goal. Through this phase the development team tries to capture the detail of each requirement and to make sure the stakeholders understand the scope of the work and how each requirement is going to be fulfilled.

b) **Design:** In this phase the development team focus on the technical architects and aspects of the intended system. The stakeholders are involved in this phase to discuss some details about various parameters attached to the technology such as potential risks, which version to be used, the capability of the team, the project constraints, time and budget are reviewed, and then the best design approach is selected for the product. This phase is very important to define all the components that need to be developed, communications with third party services, user flows and database communications as well as front-end representations and behaviour of each component.

c) **Implementation:** In this phase the real development of the technology starts by coding according to the requirements and the design discussed in the design phases. All necessary data are created in the database, and the way the intended technology is created. Also, the necessary interfaces and Graphical User Interface (GUI) to interact with the back-end all based on guidelines and procedures defined by the company are developed in this phase. This phase could take a bit long time compared to other phases as the cycle of development is repeated until the requirements are met.

d) **Testing:** This is the last phase of TDLC, where the final technology is delivered to customers. However, the development team conducts the final tests on the technology to find out defects within the technology as well as verifying whether the application behaves as expected and according to what was documented in the requirements analysis phase. If any defect is found, testers must immediately inform the developers about the details of the issue and if it is a valid defect, developers will fix and create a new version of the related software which needs to be verified again. In this phase as the previous phase, the cycle is repeated until all requirements have been tested and all the defects have been fixed and the technology is ready to be delivered or shipped to the customer.

e) **Deployment and Maintenance:** The technology is deployed to the customer for use. However, the development team or company must designate a maintenance team to look after any post-production issues. In case any issue is encountered after the technology goes live, the development team is informed to fix it immediately.

The TDLC is one of many methods can be used by management when considering a technology for business. For example, Technology Lifecycle Management (TLM) is another method and approach works through the same system having a multi-phased approach that encompasses the planning, design, acquisition, implementation, and management of all the elements comprising the IT infrastructure. However, the TLM is required in-depth technical knowledge, astute business processes, and expert engineering and financial services into a solid business model enables agencies to proactively address systematic budgeting and long-term management of their IT infrastructures.

*Case Study*: SAP ERP implementation in Bahrain Telecommunications Company (Batelco). In 2001, Batelco decided to change the stanalone system in Logistics to SAP ERP. As a member of the project team, we followed the TDLC method which extended almost two years. The project was handled by ATOS company with participation of all key users such as Logistics, Finance, Procurement, Project, and IT department. The most difficult phase of the project was "Deployment and Maintenance", as many errors and shortcomings were discovered after the system being really implemented by users, and that due to lack of full requirements that were not comperhended during the previous phases.

## 4) Technology Adoption Life Cycle (TALC)

Management is advised to be familiar with the process of adoption of new technologies and different models associated with it. There are different models developed to assess the technology adoption, such as The Gartner hype cycle (Fenn, 1995), which describes the market's initial enthusiastic response to emerging technologies, the following disappointments as those technologies face challenges in use, and the gradual understanding of the real benefits from them. Also, there is a model called "the diffusion process developed" developed by Beal, Rogers, and Bohlen (1957), which describes the process of adoption, acceptance, and eventual decline of new technological innovations at a macro, societal level. However, one of the most famous and important models that is most reliable is known as the technology adoption life cycle, which was proposed by (Moore, 1991). Geoffrey Moore described the model in his (HarperCollins Publishing, 1991) as a sociological model that describes the acceptances of new technology and product at businesses, and it is the insight into the current market conditions for a technology and a glimpse into the future. The model motivates organisations or individuals to know the commercial gain of a technology through the expense of research and development phase, and the financial return during its vital life. Such the analysis is important before the potential technology is approved by management, in order to get details about the technology in the form of its life cycle, which describes the status of technology during its R&D, Maturity, and declining stage affects the profits of it. There are five categories of psychographic profiles in the model spectrum as shown in *Figure 3.1*, namely, i) innovators, ii) early adopters, iii) early majority, iv) late majority, and v) laggards.

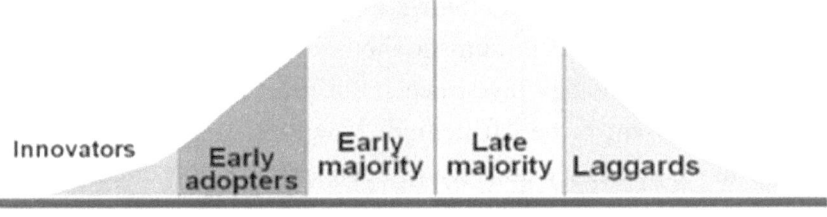

Figure 3-1 Stages in TALC

a) **Innovators:** Also known as technology enthusiasts, this group of people committed to new technology and love to get their hands on the latest gadgets. Their role is one of gatekeeper who provide access to the next segment of buyers. This group is essential to the development of new markets because they validate the usefulness of new technology. However, Innovators make up a very small percentage of the early market.

b) **Early Adopters:** Also know as 'Visionaries'. In this category, the adopters are considered the true revolutionaries in business who want to use the discontinuity of any innovation to make a break with the past and start an entirely new future. They represent the largest source of revenue in the TALC. They are believed the real competent consumers of technology, as they are willing to accept a solution that is not entirely complete, and they are prepared to commit the required resources and effort to make the technology work. From a business perspective, these people are considered very loyal to the technology they adopted.

c) **Early Majority:** Also know as 'Pragmatists'. They represent the largest source of revenue in the TALC. In this category, the bulk of all technology infrastructure is purchased for the sake of making their company's systems work effectively and look to adopt innovations only after they have established a proven track record. In other words, they are not techies, but interested in adopting a technology after being proved its usability and functionality in the market to ensure the safe purchase.

d) **Late Majority:** Also known as 'Conservatives'. The people in this category are pessimistic about their ability to gain any value from technology investments, but must adopt it because they don't want to be left behind. However, companies or people in this category are very price-sensitive, highly skeptical and doubtful, but could pay a high price under duress because the remaining alternative is to let the rest of organisations pass them by.

e) **Laggards:** Also known as 'Skeptics'. These companies or people do not adopt new technology, but could buy it only when it is buried so deep in a technology that they don't know it is there. They always depend on legacy systems until the option is no longer available. From a business perspective, they are not so much potential customers as ever-present critics, and not considered the target to the markets. Their role in the TALC is to discourage every possible application of new technologies or products.

The TALC model is applicable to Logistics because most technologies and innovations are relatively new in the logistics and supply chain. According to some studies, there are some disagreements among practitioners and experts in the logistics over which category should be approached for adopting new technologies. Some believe that the early adopter category is the best option to proceed in order to lead others in the market, and argued that being the first to try the technologies, early adopters are in a position to establish standards to their advantages. On the other hand, another group believes the early majority category is safer for the organisation, because the technologies in the early majority level are more reliable, proven, and of relatively lower price, thus the organisation is not wasting money and at the same time ensure higher levels of efficiency. This is a very critical part within the TALC model that needs a deep analysis to be conducted by organisations to understand what category is more added value to their business. In his 1991 book, Crossing the Chasm, Geoffrey Moore identified a delay in this process for radically innovative products. Moore also argued that companies with disruptive technologies faced a particularly difficult time moving from the early adopter stage to the early majority. He called the gap between these two stages "the Chasm.", and considered bridging the chasm is the key to success for high-tech startups.

> *Case Study*: Before implementing SAP ERP in Batelco, the company went through the chasm point by considering how the new technology could improve the business by taking into account the usability of SAP ERP in Bahrain Aluminum Company (ALBA). ALBA strated to use the system as the first company in the kingdom of Bahrain. After deep study, the management in Batelco convienced the early adopter stage in ALBA indicated the successful of the technology in terms of improving processe, reducing costs, and as value added to business.

## 5) Technology Acceptance Model (TAM)

Any inclusion of technologies in industries must be preceded by the user accepting the technology, because without such the effort, the technologies could remain abandoned or heavily underutilised once put into operations. Acceptance of any new technology is viewed as a function of user involvement in technology use based on a psychological process that users go through in making decisions about it. Moreover, recognising the needs and users acceptance are the beginning stage of any businesses and this sympathetic would be supportive to find the way of future development, thus researchers and practitioners are interested to realise the factors that drive the users' acceptance or rejection of technologies. For that reason, many studies have been conducted to realise the value of a new technology from the users' perspective, and accordingly, a number of models and frameworks have been developed to explain user adoption of new technologies and these models introduce factors that can affect the user acceptance. The TAM model is one of various technology acceptance frameworks were developed for predicting the relationship between a technology and user's acceptance. To come to a better understanding of the situation surrounding the development of the TAM, it is important to have a concise narrative of theories and models preceding its manifestation. For instance, Theory of Reasoned Action (TRA) was developed to predict and comprehend human behaviour and attitudes, The Theory of Planned Behaviour (TPB) was formulated to take care of the limitations of the Theory of Reasoned Action (TRA) and set out to predict the intention of people to engage in behaviour within a particular place and time and to describe all behaviours over which an individual has the capacity to apply self-control. To this end, Davis (1989) modified the above theories and originated the TAM which aims to predict the acceptance and rejection of modern technology. David stated the goal of TAM as:

*"To provide an explanation of the determinants of computer acceptance that is general, capable of explaining using behavior across a broad range of end-user computing technologies and user populations, while at the same time being both parsimonious and theoretically justified"* (Davis 1989).

Technology Acceptance Model (TAM) was the first model to deal with the psychological factors affecting technology acceptance through a simple model that is considered easy to use by many researchers and practitioners for predicting the user acceptance any new technology being adopted by organisations, and over the years, TAM2 and TAM3 have been developed by the original authors to include more external variables based on the type of technology being adopted and its effect on a business's performance for each organisation. One important reason that businesses need to consider the TAM model is the mandatory of using IT/IS in organisations, which means users must use the system for certain transactions, and users must accept any technology being adopted in order to ensure the future usability and the potential for vastly improving the performance of individuals and organisations. Moreover, the TAM as shown in ***Figure 3.1***, is a parsimonious model simply because it is composed of three predecessor variables (Perceived Usefulness (PU), Percieved Easy of Use (PEOU), and Behaviour Intention (BI)). The two main keys in the model are PU and PEOU refer to; PU is a person's degree of belief that using a technological system will increase his/her job performance, and PEOU is a person's degree of belief that using a technological system requires no additional effort, which are extrinsically factors that motivate users' acceptance, adoption and usage behaviour of a technology. Further, The two factors are considered to be crucial to the study of user acceptance as they are directly or indirectly influence actual system usage through behavioral intention. In other words, behavioral intention is jointly determined by perceived the usefulness and attitude of users. Perceived ease of use has a small but significant impact on perceived usefulness, and both have an impact on the perceived attitude to use a new technology. TAM is one of the most credible models compared to other models according to some studies conducted in the last 20 years[9], and proved its credibility in terms of explaining the rate of acceptance at the time of introducing the new technology. Also, TAM has been used in many studies related to technologies and systems, such as email, voice mail, graphic, spreadsheet, e-Commerce, and others. The model has proven as a powerful tool to explain technology adoption in terms of time, subjects, context, and it allowed around 86 models to provide a good explanation for a technology, contexts and expertise level.

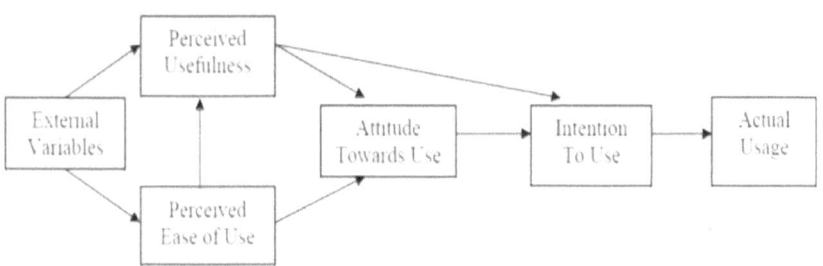

Figure 3-2 First version of TAM (Davis et al., 1989)

In the logistics context, it is very essential to understand information systems and technologies that used for day-to-day operations, and without such the information the whole business operations could be ceased or suspended. In today's business environment logistics services

---

[9] Using the TAM2 model for investigating e-Government Adoption in Bahrain and evaluate the key determining factors for strategic advantage, by Dr. Ali Kamali. PhD Thesis, LSBU, 2018.

have become a vital tool in gaining a competitive advantage, and Information technology (IT) is considered the competitive weapon and determinants of organisation's competitive power, thus the behavioral intentions of employees to use new technologies is important to be investigated to help organisations scale up. The operations in the logistics are a set of approaches collaborating important participants such as suppliers, manufacturers, distributors, and stores, and information technology is the rope that ties together all those important nodes. With growing technologies in the logistics operations, the complexity and misunderstanding of work with a new system can add additional stress on operators (users), and inaccurate prediction is likely to have worse consequences for the business. Barriers such as resistance to change from employees, employee skills and knowledge shortage were proven as significant factors in testing when using information systems in logistics. There are several advantages for investigation user's acceptance in the logistics such as:

i) Helps management to help an individual's ability to perform his or her job better.
ii) Ensures the success of technology adoption and the positive results of investment into IT projects in organisations.
iii) Explains technology acceptance to be relevant to address the crosscutting issues among functions.
iv) Identifies a valuable tool in forecasting satisfaction, improving customer service and improving service quality.
v) Provides a basis for identifying the influence of internal beliefs, attitudes, and intentions and how these can be manipulated.
vi) Develops several hypotheses as part of a research study through proposing relationships between the variables need to be investigated.
vii) Helps to understand and explain use behaviour in information system implementation.
viii) Offers a basic framework to explain the influence of external variables towards.

ix) Predicts IT acceptance under different conditions, such as time and culture, with different control factors.
x) Explains technology adoption in a wide variety of contexts, ranging from individual to organisational technology acceptance.

# PART TWO

# Components of Material Management

# CHAPTER 4

## The Warehouse

### 1) Introduction

Warehousing is considered the key part of Logistics and Supply Chain through its determination and ability to establish smooth and efficient logistic operations in organisations, and such operations play a vital role in determining a company's competitiveness, as logistic costs are considered an important part of the overall production costs. The warehouse is the building where the core materials for business such as raw material, work in progress, finished goods, and parts are stored in. One important thing must be mentioned here that a warehouse in any organisation is only utilised for storing stock items, not non-stock items. The difference between a stock item and non-stock items that the former refers to those items that are kept physically in a warehouse for operations and sales, and they are restrictively controlled in the inventory system, where the latter refers to those items that do not keep physically in a warehouse. In other words, non-stock items that are not subject to tracking by a perpetual inventory computer system and do not have an item master record or an SKU number in the database, and the business does not sell them but use them for internal operations e.g. cleaning tools, stationery items for employee's daily work, etc.

Basically, businesses think of warehouse in advance, and allocate the necessary budget, which is normally high, to ensure achievement the optimisation process and performance improvements, the warehouse design is considered one of highly complex tasks and takes a high part of the business budget.

The main purpose of a warehouse is to cover activities related to supply management through inflows and outflows of stock items in an organisation. Also, it can be identified as the main part that make up the good customer service as the goods transported from the manufacturer to the warehouse and then to the final recipient (customer/consumer). To this end, if a warehouse is not effectively managed the whole supply chain may be paralysed, through stock unavailability, slow or delay the movement of the items, and ultimately customer dissatisfaction. The warehouse is also considered as a spot for generating huge costs for organisations in terms of inventory storage costs typically include the cost of building, material handling equipments, IT hardware and applications, day to day operations, and human resources employed to run the warehouse.

## 2) Scope of Activities in a Warehouse

The main scope of any warehouse are namely: stock movement, storage and management of stock, and information flow.

a) **Stock Movement:** The activity concerns with receiving new or materials from suppliers, putting away stock processes, transferring materials to Production, and shipping finished goods to customers. Beside, controlling inspection and quality processes and packaging.

b) **Storage and Inventory Management:** This activity concerns with having the right inventory at the right quantity, in the right place, at the right time, and at the right cost in the warehouse. Managing stock must be efficiently implemented by the store/warehouse manager to achieve economies of scale in production, transportation and handling of goods. The way to managing inventory largely depends on the type of industry and product. For example, the storage of oil and gas products is more complex than storing perishable goods or products in terms of storage, delivery, packaging, handling equipment, inventory counts, and safety aspects.

c) **Information Transfer:** The third activity is information transfer, as it occurs simultaneously with the movement and storage activities. Every movement of materials must be related to the transfer of information to avoid any uncertainty regarding the stock movements and management. Sharing information enables the seamless flow of transactions among all stakeholders in the logistics context, and can obviously be a way to coordinate effectively with other functions in the organisation such as Finance, Purchase, and Production, and externally with Suppliers and Distribution centres in terms of lead times and transportation methods, because it is the only way to ensure full integration of all elements of material flow management, efficiency and reliability of their interaction. With advanced technology, the integration of information processes is interrelated and interact through a sharing database internally and externally (e.g. Oracle used ERP).

## 3) Functions of Warehousing

Functions of Warehousing cover up the wide ranges of activities which are associated with the main scope of activities mentioned in the previous section. The main roles of warehousing in most organisations, especially in the manufacturing industry, physical distribution of goods from end of production line to the final consumers. In commercial industries, the distribution of goods takes a different way as goods move from suppliers to end customers. These activities include purchasing of goods, inventory management, storage, materials handling, protective packing and transportation. Also, the warehouse takes responsibility over the safe keeping of goods until they are needed for consumption. The following are the main functions of warehousing:

a) **Storage of Stock Items:** This is the basic functions of any warehousing to provide the facility of storage for the goods which are laid with organisations. A warehouse acts as a storage space for all types of stock items, using scientific methods in making them available easily and smoothly when needed. Inventory warehouses often contain a large number of shelving units or storage containers and may feature a computerised inventory tracking system to assist with keeping track of items contained in the warehouse. In all warehouses, there are two main functions; inbound functions assist to prepare for storage as well as outbound functions pack and ship orders, resulting in benefits for both the business and customers.

b) **Receiving of Materials:** The receiving section of a warehouse is the area where all supplies that are either imported or locally produced are received. One of the important functions of the receiving section is to check items delivered to the organisations, either coming in as new stock or as supplies. This includes inspecting the quality, condition, and quantity of any incoming goods, and allocating them to a space in the warehouse. The receiver(s) is responsible to conduct an exhaustive analysis of the received items to ensure the delivery has come to the right place, right quality, right quantity, and right specification/description of item by matching the details on the consignment/delivery note to the purchase order (PO) raised by the purchasing department. It is important for the receiver to ensure the received items or their packaging are not damaged, broken or malfunctioning before raising a goods receive note (GRN) in the system (e.g. ERP or/and WMS). Once the item is approved, the receiver either implements a direct putaway that to store the item in its designated location (recorded location in the system), or using a chaotic putaway by putting the item in the first available spot then assign a logical location for it in the system.

c) **Quality Control or Quality Inspection:** This is one of the most essential core component of any good warehouse to ensure efficiency for holding items. This function is a vital aspect of stock control, especially as it may affect the safety of customers or the quality of the finished product, and to verify that the product is within certain prescribed tolerances in order for the product to be useful. However, the quality control function is not always set under the warehouse management, but could be assigned under the production management, especially in the manufacturing companies. The main role of the quality control/inspection is to perform a quality inspection for any purchased items from suppliers. Also, the function is responsible to conduct inspection on products during the manufacturing process, inspection of the final finished product, inspections while the items are stored in the warehouse, and inspection before the item being loaded for dispatching, including the packaging used to ship it to the customer. The quality control function is a prerequisite to achieve relevant ISO certifications, which offers customers a greater sense of expected quality and may lessen the requirement of regular inspections

d) **Physical Stock Count and Cycle Count:** According to Generally Accepted Accounting Principles (GAAP) and Internal Revenue Service (IRS) rules, businesses must conduct an inventory count periodically for items being stored in a warehouse. This is one of warehouse's responsibilities to ensure stock accuracy and determine any mistakes might affect on the accuracy of inventories in stock. This function is the actual count of the materials and products a business owns. There are two different processes and methods to implement this function and that are decided based on each organisation's policy; Cycle Counting or Physical Inventory, or/and might be together. The cycle counting process is an inventory auditing procedure that is a part of an inventory management solution, where a small subset of inventory in a specific location is counted on a specific day. The process has some advantages that enables organisations to count multiple items within the warehouse, without needing to count the entire inventory, and if any error discovered or uncovered in stock item in terms of inaccuracy compared to inventory records the warehouse can correct it immediately, to ensure the physical stocks reflect the stock record's balance. Whereas the physical count process is an in-depth count and it is usually conducted once a year, or with smaller sample counts several times in the year (i.e. quarterly or twice a year). However, the cycle count process is more cost-effective solution than the physical count, as it avoids interruptions in operations with less complexity, since the physical count is normally conducted to cover up all stock items in the warehouse, thus whole operations must be stopped for one day or more which can have a major impact on the business. Nevertheless, conducting any method depends mainly on the amount of inventory in each organisation. In other words, for organisations with a smaller inventory, conducting the physical count is a better option since it allows the organisation to start the new year with a clean slate, ensures a more accurate count of products, and maintains focus while keeping inventory as a priority, but if there is a large amount of stock, the cycle count method will be better to cover up all stock items through daily stock counts for some number of stock item throughout the fiscal year.

> *Case Study*: Deeko Bahrain is a manufacturing company in packaging and disposable products used to conduct the stock count every quartel due to huge stock items in different warehouses. When I was managing the warehouses as Head of Warehouse and Logistics, I had to stop all transactions throughout the period of stock counts which sounded difficult for the production, the sales, and the finince department. For that reason, the management decided to appoint a special employee to count stocks daily throughout the years in order to prevent such difficulties which used to occure through every periodical stock counts process.

e) **Managing Material Handling Equipment:** This function is very important to ensure the efficiency and speed of warehouse operation, which ultimately result in elongated order completion cycles, and for that reason the decision for investing in material handling system is strategic in nature to add value to the business. The warehouse is responsible to suggest the most appropriate materials handling equipment for daily operations, giving careful attention to the safety aspects and business requirements at the minimum possible overall cost. The decision to acquire any material handling equipment must be related to the movement, storage, control and protection of materials, goods and products throughout the process of manufacturing, distribution, and consumption. In other words, material handling equipment must meet the purpose of the key activities in a warehouse such as loading, unloading, palletising, and de-palletising. Also, the warehouse management is responsible to maintain material-handling equipment based on a preventive maintenance method through regularly inspecting equipment, systematically conducting maintenance tasks and correcting issues as soon as being discovered, rather than waiting until they become major failures, and that is normally conducted through an agreement with a professional provider of preventive maintenance services, but some organisations prefer to have their own professional to look after the maintenance aspect to save cost and time. Another warehouse's role in this respect is to select operators who are qualified for performing the job, along with regular trainings arrangements to enable them to have a good understanding about the safe handling of materials, goods and equipment during operations.

f) **Dispatch of Orders:** This is one of critical functions in a warehouse as it related to client satisfaction, and thus it must be properly planned and controlled. The main role of the warehouse manager in this respect is to fulfill the delivery commitment based on the agreed time, which is known as the delivery date for dispatching orders to the customers.

The warehouse is responsible to ensure the quantity, right quality, right time, and the right transportation method before dispatching items to the customer. They also pay attention to the rout management to ensure dispatch efficiency, time efficiency, and road safety for trucks being used in daily operation, and to adjust and respond to the day-to-day realities once vehicles leave the warehouse and execute on the road, and the route performance is measured against plan on a continuous basis to achieve maximum productivity, meet customer expectations, and support continuous improvement. Also, the warehouse is responsible to coordinate drivers and dispatchers to streamline operations.

g) **Information Management:** Managing information is the core activity to make all other processes efficient and achievable in a warehouse. Therefore, the warehouse is responsible to ensure all transactions are recorded and updated on a real-time basis to reflect the accuracy of information in the system for the firm operational process performance. Any information being recorded in ERP or/and WMS is shared with other departments in the same organisation or with external stakeholders such as suppliers and distribution centers, and thus can leverage real-time information to gain insights into customer satisfaction and workforce performance. The warehouse's role is to manage all transactional data and records in receiving, dispatching, picking and putaway, bin locations, and physical count processes, in order to maximise productivity, reduce errors, control the processes, and thus become an effective function within the logistics cycle. As a valuable consequence, effective contribution in improving organisational agility, better and faster decision making, quicker problem-solving, better communication, and improving business processes. Finally, the warehouse is responsible to ensure the throughput is processed smoothly at the speed at which products flow in the system could be evaluated, and regular checks are conducted to assess an impact on the speed of the system.

## 4) Warehouse Layout

The layout of a warehouse is an important as it includes a huge investment and budget to cover various costs such as the rent, maintenance cost, inventory cost, storage facilities for products, and the decision should be based on how an organisation is able to shell out to maintain the warehouse functions, and without investing in optimisation of the warehouse layout the business might become in a suboptimal situation with high warehousing and handling costs, less efficient processes, and subpar customer service. Therefore, management must work rationally to set the objectives in advance to avoid barriers in the future.

The warehouse should be designed in a way to work with a space in which certain factors limit the surface area available, and to cover all requirements for day to day activities in the warehouse. The layout must be built to cover the following needs in the warehouse:

i) Loading and unloading areas.
ii) Reception area for new receipts.
iii) Storage area for storing items.
iv) Picking area for assembly.
v) Dispatch area for delivery.

To achieve the maximum efficiency and space utilisation, organisation should consider the following factors when deciding to build a warehouse:

i) Making the most of the available space
ii) Reducing the handling of goods to a minimum
iii) Providing easy access to the stored product
iv) Having the highest rotation ratio possible
v) Offering maximum flexibility in the positioning of products
vi) Controlling the amounts stored

The layout as shown in ***Figure 4.1*** should be built in a way to smooth inbound and outbound of materials, such as Docks, Picking storage and Racking systems for storing items. Furthermore, there should be the capacity in the building for pallets which are normally put on the racking system inside the warehouse, considering the size of the same. The business should accurately analyse how many docks should be set up for loading and unloading goods, and that must be based on the amount of business in terms of holding stock, transactions, customers, and supplies, and consideration should be taken for scalable processes to avoid languishing in operations in the future. For example, organisation should have a plan to accommodate shifts in business, inventory, or process changes, thus such plans should be prepared in advance when installing a new warehouse.

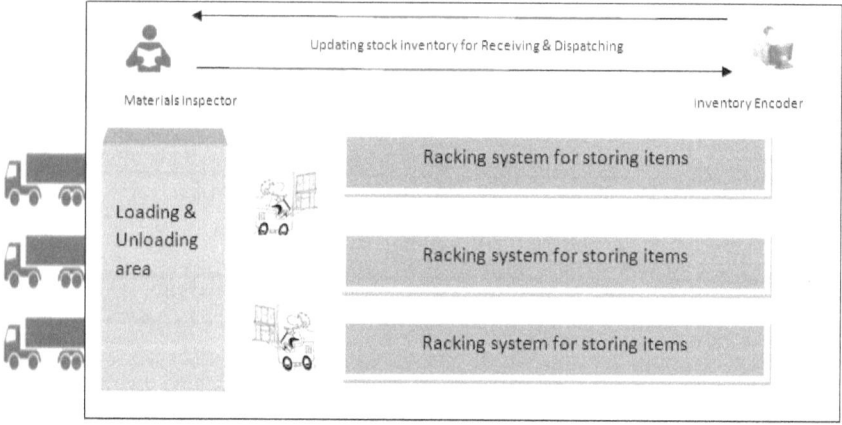

Figure 4-1 The basic layout for a Warehouse

## Racking Systems

Attention must be paid to maximise space efficiency through installing a pallet racking system to optimum throughput efficiency, but it should be designed in a way that making the best possible use of the available floor and vertical space in a potential warehouse. However, the type of racking system can't be decided if the building for which the storage system is to be designed has not been built, as the key factor in selecting the racking system depends on some basic requirements such as pallet load dimensions, forklift aisle requirements, and the roof-support column locations. Nowadays, most racking systems are flexible and adjustable for dealing with changes of loads and new business requirements. In other words, adjustable racking systems can be set up in a variety of ways to meet organisation's operational needs, the size of the facility, different handling equipment used, and for the required pallet load. There are different types of racking systems available for a business as listed below:

a) **Cantilever Racking System:** this type is used for storing long products, such as lengths of timber, plastic piping and steel rods, and consists of long arms protruding from a metal framework specifically designed to store long or bulky items. *Figure 4.2* shows Cantilever Racking System.

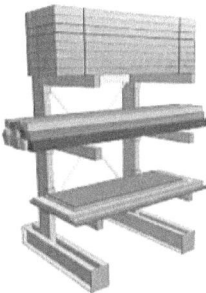

Figure 4-2 Cantilever Racking System

b) **Selective Racking System:** This type is considered the most popular as it allows direct access to each pallet and can be configured to almost any size business desires, and it can be easily installed comparing to other types. *Figure 4.3* shows Selective Racking System.

Figure 4-3 Selective Racking System

c) **Push Back Rack System:** This type is mainly used by organisation conducts the LIFO method. This type allows for specific SKUs, and works through pushing the next pallet on the rails where it rests firmly when a new pallet is loaded on the structure. When unloaded, they are 'pushed' to the front of the structure. *Figure 4.4* shows Push Back Rack System.

Figure 4-4 Push Back Racking System

d) **Drive-in Racking System:** This system is designed to maximise the use of floor space within a warehouse and has a LIFO rotation. This type of pallet racking requires fewer aisles for the same amount of storage. Hence, because of the space saved, they are a cost-effective option. Drive-in racking is designed in a way which allows forklifts to make their way through the aisles to continue stocking and adding additional pallets. *Figure 4.5* shows Drive-in Racking System.

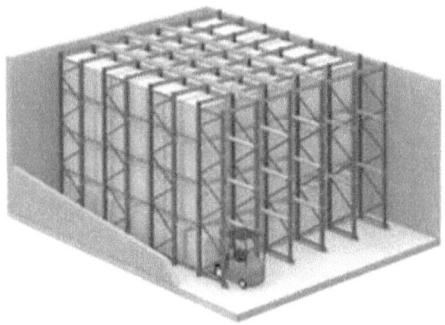

Figure 4-5 Drive-in Racking System

e) **Pallet Flow Racking System:** This type is applicable to the FIFO method. All loads are stored at the higher end and removed at the lower end point. As products continue to be loaded, the rotation becomes automatic because of the flow of the racks. This type is typically ideal for high density storage with multiple order picking levels. *Figure 4.6* shows Pallet Flow Racking System.

Figure 4-6 Pallet Flow Racking System

f) **Carton Flow Racking System:** This type is also applicable to the FIFO method, as the system uses a rear load design, which allows for easily operated and managed inventory. Unlike racking systems, carton flow racking rotates products automatically by design, allowing for storage optimisation and maximum efficiency. *Figure 4.7* shows Carton Flow Racking System.

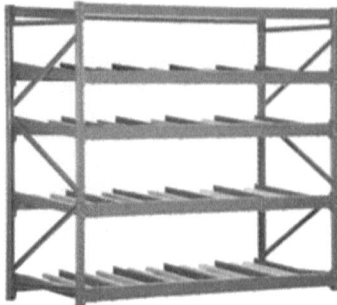

Figure 4-7 Carton Flow Racking System

## Key factors in the warehouse layout

i) In a manufacturing industry, warehouses that provide stock to manufacturing operations must place raw materials and packaging close to the production area, in such a way that the production floor's demands are easily met.

ii) The roof must be designed to keep out the weather.

iii) Think of ventilation requirements and type of air conditioning system, especially for perishable items.

iv) Attention must be paid for upwards than outwards (per cubic meter of space).

v) Think of the amount of activity that is to take place and ensure that adequate movement space is also provided.

vi) Unitise goods for ease of handling and storage.

vii) Use industry-standard pallets and containers to eliminate double-handling, minimise the risk of product damage and make best use of space.

viii) Select the right material handling systems for holding, transferring or storing materials over a period of time.

ix) Prioritise safety and health of staff, products, and facilities when planning for a warehouse storage and design. Sprinkler systems and fire extinguishers must be highly considered with regard to safety aspects, and regular tests must be conducted to ensure their functionality.

x) Think of security aspects in a warehouse such as tight round the clock security, surveillance cameras, and security guards.
xi) Choose the right information system that provides the right solutions for the organisation's needs.
xii) Think of safe and protected storage of hazardous materials with clear procedures for the way of handling.
xiii) Think about areas for forklift battery charging in the warehouse.
xiv) Conduct continued training for staff members to improve efficiency, in order to ensure the operations of the warehouse run smoothly and in the correct way.

## 5) Leasing vs. Buying a Warehouse

This is a crucial stage for any organisation to decide between buying a warehouse property or finding a warehouse space for rent, and such decision should be carefully examined based on several factors such as the anticipated capital requirements of the business, the amount of business, and profit vs. loss as outcomes. However, both of these two options have advantages and disadvantages, and for that reason some organisations prefer to approach consulting firms before taking the final decision about the best option that could add value to their businesses.

### Renting a Warehouse Facility

**Pros:**

i) More options available when it comes to choosing location, size, price, and other important amenities.
ii) No down payment is required, and instead the money can be invested in other areas of business.
iii) The organisation doesn't bear the costs of repairs and maintenance around the facility. However, it depends on the type of contract between the two parties.

iv) It gives more choices in the size and type of warehouse required by the business.
v) It saves management time, and thus they are able to focus on the core aspects of business instead of doing the behind-the-scenes aspects of running a warehouse.
vi) It is a better option if the business at the starting point to evaluate to what extent a warehouse could add value to the business.

**Cons:**

i) Commercial leases are generally subject to annual rent increases as well as increasing costs whenever the lease is renewed. However, increase the costs depends on the agreement between the parties, and so that the picture is clear for all of them in advance.
ii) There might be control loss over the facility, and reduction in management's overall ability to control the decisions of the property.
iii) The option is not practical if the business is projected to expand and grow fast as the modification the current facility may cost the business a lot for relocation.

**Buying a Warehouse Facility**

**Pros:**

i) A fixed cost ensures the monthly rate remains more consistent over time.
ii) The tax deduction can be claimed with this option for all costs associated with owning, maintaining, and running the facility.
iii) Using mortgages loans have less rate fluctuation than rental rates do, ensuring consistent monthly rate comparing to lease payment over time.

iv) Better control over the property as it gives the owner more guarantee in business dealing with clients.
v) Equity from the growth of business can be used as leverage for obtaining loans, funding, and other business dealings.
vi) Better control over the facility so it could reduce anxiety from any intervention from the landlord.

**Cons:**

i) It is required high upfront costs for paying for the property and equipment.
ii) It increases the financial risk through increasing the costs of tax and insurance.
iii) In case of agility, the business faces some difficulty to move quickly, especially if demand is low when selling the property.
iv) Maintenance costs are high for having total control over the building.
v) Not practical in case the business grows and hence the premises might be inadequate.

## 6) Stock Control Techniques in Warehousing

Inventory control techniques are the most important methods to integrate all aspects of an organisation's inventory tasks, including procurement, distribution, and warehousing. In other words, controlling stocks is the top priority for all stakeholders in an organisation such as the warehouse manager, production manager, purchasing manager, and of course the finance manager whose role to ensure the stock balance which results is an investment that needs financing. The stock control techniques or methods are employed to ensure the balance of stock held between supply and demand on one side and costs on the other, and such methods are important to ensure the optimum quantities of stock items at all stages in the production cycle and evolving techniques which would ensure the availability of planned inventories. For example, carrying too much stock items might leave an organisation with a larger dollar investment and increased risk if a product loses favor in the market. Conversely, holding too little stock on hand can create a shortage and make the business miss out an opportunity to cash in when consumers are ready to spend. To this end, several techniques were developed for controlling stock in warehousing to provide an efficient system to optimise productivity and efficiency along the logistics while having the right inventory at the right locations to meet customer expectations, and new programs can calculate all methods automatically. Importantly, both obsolete and non-stock items must not be taken into account when applying any techniques. The following are the most common techniques are in use by warehousing to control stock items:

a) **ABC Analysis**

   The analysis is to define an inventory categorisation technique often used in Materials Management, and Flores and Whybark (1987) who first proposed the technique. This technique is based on the Pareto principle for determining which items should get priority in the management, and it is usually categorized items to three classes/codes as follows:

   "*A class*" stock items that account for 80% of total value, or 20% of total items.

   "*B class*" stock items that represent around 15% of total value, or 30% of total items.

   "*C class*" for the remaining stock items, which typically represent 5%, or 50% of the total items.

   The formula to identify the three categories is calculated as follows:

   > Item cost*Quantity issued/Consumed in period

## Advantages of ABC Technique

i) **Minimum Ordering Cost:** Under ABC analysis, the materials from group 'A' are purchased in lower quantities as much as possible. With this, the effort to reduce the delivery period is also made. These in turn help to reduce the investment in materials.

ii) **Strict control:** Under ABC analysis, strict control can be exercised to the materials in group 'A' that have higher value.

iii) **Minimum Holding cost:** Since the materials from group 'A' are purchased in lower quantities as much as possible, it reduces the storage cost as well.

iv) **Saving in time:** Since a significant effort is made for management of the material from group 'A', it helps to save time as well.

v) **Economy:** This method is economical, since equal time and labor is not needed for all types of materials.

vi) **Flexible:** It is very easy and flexible to apply in order to benefit from space in a warehouse

b) **XYZ Analysis**

This technique is almost similar to the ABC in terms of holding the most expensive items in stock. However, the difference is that the items are first sorted in descending order of their current stock value. The values are then accumulated till values reach say 60% of the total stock value. These items are grouped as 'X'. Similarly, other items are grouped as 'Y' and 'Z' items based on their accumulated value reaching another 30% & 10% respectively. The formula is used to identify the highest value item:

> Item's current stock value / The total stock value in the warehouse

The XYZ analysis gives an immediate view of which items are expensive to hold. Through this analysis, organisation can reduce its money locked up by keeping as little as possible of these expensive items.

c) **Economic Order Quantity (EOQ)**

This is one of the oldest techniques developed to propose the ideal order quantity a company should purchase for its inventory given a set cost of production, demand rate and other variables. The technique was developed by (F. W. Harris in, 1913). The technique is done to minimise variable inventory costs, and the equation for EOQ takes into account storage, ordering costs and shortage costs. The formula is used to calculate the EOQ as follows:

$$EOQ = \sqrt{\left(\frac{2SD}{H}\right)}$$

Where:
S = Setup cost (per order, generally including shipping and handling)
D = Demand rate ( Quantity sold per year)
H = Holding cost ( per year, per unit)

The EOQ technique requires an efficient inventory system to maximise profit thus, an organisation is able to determine the optimal number of units of the product to order so that it minimise the total cost associated with the purchase, delivery and storage of the product. In other words, the formula calculates the most economical number of items a business should order to minimise costs and maximise value when re-stocking inventory. However, the EOQ requires very accurate data such as the details of order costs and inventory carrying costs to come out with suitable results. Also, some important factors must be taken into consideration when determining the EOQ, and without having accurate details of such factors the model can't practically work out. For instance, demand is assumed steady throughout the year and that inventory gets used at a fixed rate, the lead time remains the same for each order, and each unit has a cost for warehousing and holding it until being ordered again.

**Advantages of EOQ**

i) The technique is very insensitive to parameter errors, imprecise estimates, forecast, costs, because EOQs can be rounded off without a significant loss in economies.
ii) It is very practical for businesses that have multiple orders, specific release dates for their products and have requirements plan for their components.
iii) The technique provides specific numbers particular to the business regarding how much inventory to hold, when to re-order it and how many items to order.
iv) It helps to determine the exact quantities needed for the business.
v) It is reliably and accurate that it can help a warehouse run smoother and more lean by ensuring high standards of customer service while reducing overall stock costs.
vi) Calculating the formula is fairly simple using EOQ, provided the correct data inputs.
vii) The technique is useful for repetitive purchase, maintenance, repair, and operations (MRO) inventory, and could be used for seasonal stock as well, with availability of correct historical date.
viii) It is one of the most successful methods for balancing the costs of inventory with the benefits of inventory.

d) **Material Requirement Planning (MRP)**

Material requirements planning (MRP), which was developed by Joseph Orlicky in1964, is an information system in which sales based on forecasting are converted directly into loads on the facility by sub-unit and time period. In other words, the technique is a push type of inventory control, meaning that organisations use forecasting to determine the customer demand for products. The technique is used to support the planning of material requirements on all manufacturing levels to ensure to have materials, which are named as dependent demand, at the right time, right quantity, and the right place. In short, the technique is designed to answer three questions: i) What is needed? ii) When is it needed? iii) How much is it needed?

The system is typically implemented in the manufacturing industry to maintain a smooth material management process based on formulations, technology, orders from customers and to suppliers, inventory, and sales projections. When there is any shortage of the demand such as a raw material, semi-finished or finished product, the system alerts the warehouse and offers supply suggestions. It is important to note that MRP only works with a computer-based system, through developing data, which is the relationship between the materials and the finished product known as bill of material, to organise the timing and ordering of the dependent demand products, and the system then calculates the demand for the raw material and components of the final product using the demand for the final product and it is determined how much and in what quantity to order from these components and raw material, considering the production and lead times and counting back from the delivery time of the product. Furthermore, the data structures used in the applied system are divided into two categories; master data and transaction data. The former are data that exist independently of specific orders (customer, inventory, forcast, production, purchase, and transport orders), and it comprises the frame in which the planning and controlling of orders takes place. The latter is created based on the information gathered and scheduled through the master data to enable material planner to create a material plan and then to send the plan to procurement to initiate a purchase order from a supplier.

A typical calculation of MRP performed by the system is as follows:

---

**Required Quantity** = From bills of material

**Consumption during lead time** = ( ( 3 month required qty ) /90 ) x lead time

**Safety stock** = ( ( 3 month required qty ) /90 ) x Safety stock days

**If Order Quantity to Specific Size :**

**Order Quantity** = ( Material required / order size) = No. of lots rounded to higher side

**Order Quantity** = No. of lots x Lot size

**If Order Quantity up-to certain level :**

**Order Quantity** = Match Material required with min. order size , if required size is higher then order else order min. order size

---

The technique was developed over time to include more parameters and data. In 1983, Oliver Wight took MRP a step further and developed manufacturing resource planning (MRP II) by including two more data i.e. employee and finance information. In 2011, the new version of MRP was proposed by by Joseph Orlicky's original publishing company, McGraw Hill. They introduced demand-driven MRP (DDMRP), with five components, including strategic inventory positioning, buffer profiles, level, dynamic adjustments, demand-driven planning, and highly visible and collaborative execution.

## Advantages of MRP

i) Allows low levels of in-process stocks.
ii) Allows to track the component needs directly from sales orders.

iii) Allows to evaluate the capacity requirements suggested by the main schedule.
iv) Allows to monitor the minimum amounts for each warehouse.
v) Allows to know the time required to order and deliver goods.
vi) Reduces delivery and ordering costs by stacking together all orders to one supplier.
vii) Allows the warehouse to automatically create delivery and economic orders.

e) **Just In Time (JIT)**

The Just in Time or JIT method is one effective solution to reduce costs, improve quality and meet the ever-changing customer needs. The philosophy, which was initially used by Ford Motor Company in the early 1920's, emphasises on increase productivity in warehousing as a significant method in reductions in inventory, and thus several companies have adopted the approach as their primary inventory management technique as it works through decreasing total inventory levels on a continues basis by decreasing the lot size, and thus the amounts of buffer stocks, work in process, and in-plant inventory. Moreover, the JIT system involves many functional areas of a firm as well, such as manufacturing, engineering, marketing and procurement, among others. Essentially, JIT is the method implemented to eliminate the waste by producing and delivering finished goods Just-In-Time to be sold, sub-assemblies Just-In-Time to be assembled into finished goods, fabricated parts Just-In-Time to go into sub-assemblies, and purchased materials just-in- time to be transformed into fabricated parts. In other words, JIT disseminates the idea of having the right material at the right time, the right amount, and in the right place. However, JIT can't be implemented efficiently without an effective relationship between the buyer and the supplier. In other words, a long-term partnership and continuous communication between them will eliminate the need for frequent re-bidding, since every supplier is informed of the production planning, material needed, delivery arrangements, and the related purchasing requirements. Furthermore, through a good relation with the supplier, the company can ensure working with short lead-time since the whole procurement cycle is coordinated so well. Therefore, it is one of the advantages of working with suppliers in this way.

## Advantages of JIT

i) **Reduction of lead-time:** Will result in higher levels of customer service and lower safety stock requirements for the company. Lower levels of safety stock contribute to reduced working capital requirements for the company.

ii) **Elimination of wastage from the processes:** Will improve the quality of materials, semi-finished and finished product, and hence it contributes to eliminating wastage from processes.

iii) **Reduction of non-value-added activities:** Will result in enhancing the whole activities in Material Management and then non-value-added operations can easily be identified and hence eliminated.

iv) **Reduction of inventory holding costs:** Will result in minimising the WIP inventory, raw material and the finished goods.

v) **Close relation with suppliers:** Will help to have a very strong and long-term relation with suppliers.

vi) **Reduction of the time of manufacturing process**: This is because the required materials with the required quality will arrive at the required time.

## 7) Material Handling Equipment (MHE) in a Warehouse

There are different types of MHE being used in warehousing from apparatuses and storage units to vehicles and real machines. A warehouse can't be managed without MHE, which deals with transportation, storage, control, stock counts, and protecting products or materials during manufacturing, distribution, consumption or disposal in daily basis. Further, MHE's main role is to handle a wide range of equipment without making any physical wounds to workers, and at the same time performs the job faster than the manual processes.

## Types of Material Handling Equipment

a) **Transport Equipment:** Any MHE that is used to move material from one location to another (e.g. between workplaces, between a loading dock and a storage area, etc.) while positioning equipment is used to manipulate material at a single location.

**Example:**
Forklifts, Conveyors, and Cranes

b) **Engineered Systems:** These types are used to handle material at a single location to enable storage or transport. Systems are often automated and can be used by a worker or can perform fully automated functions. Such systems are considered the new generation of material handling systems and they are the most widely used today in Flexible Manufacturing Systems (FMS) and Computer Integrated Manufacturing (CIM).

**Example:**
Bulk material handling, Conveyor system, Robotic systems (Details about each one in Part 3).

c) **Bulk Material Handling:** These types are typically used to control loose materials in bulk forms such as food or liquid. Equipment such as conveyor belts or elevators that are designed to move large quantities of material in loose form, or in packaged form, through the use of drums and hoppers.

**Example:**
Conveyor belts, Stackers, Reclaimers, Bucket elevators, Grain elevators, Hoppers, Silos.

d) **Storage Equipment:** These types refer to non-automated equipment used for holding or buffering materials while not in use or in transport. The main function of storage equipment is to minimise handling costs by making the material easily accessible and gaining the maximum use of space. Such systems have a variety of characteristics to make them suitable for different operations. Some trucks have forks, as in a forklift, or a flat surface with which to lift items, while some trucks require a separate piece of equipment for loading. Trucks can also be manual or powered lift and operation can be walked or ride, requiring a user to manually push them or to ride along in the truck. A stack truck can be used to stack items, while a non-stack truck is typically used for transportation and not for loading.

**Example:**
Racks, Pallets, Shelving, Storage drawers, Mezzanines, Stacking frames.

# CHAPTER 5

# Procurement (Purchasing)

## 1) Intruduction

As stated by Donald Waters in his book *"Procurement and purchasing are often taken to mean the same thing. Usually, though, purchasing refers to the actual buying, while procurement has a boarder meaning."* (Waters 2003). So the two words are interchangeable in most, when they are used as the definition of the function in an organisation. The procurement function has extremely progressed in recent years and many organisations have changed their views about its roles from the traditional view to the strategic view. From the traditional view, the procurement used to be considered as one of a standalone function where activity is confined to receiving purchase requests from internal users and translating these into purchase orders or other contractual relationships with suppliers. It was also seen as a paperwork intensive clerical function which focuses only on transaction processing such as negotiating with suppliers, selecting the right materials, at the right time, in the right quantity, from the right source, at the right price. However, the new trends in Procurement push the function to think outside the box by making it more creative and innovative, and that due to current competitive markets that require the procurement to have the ability to respond at speed by focusing on making buys feel fast. To achieve that, organisations have changed their tactics by increasing the roles and responsibilities of the procurement function to think strategically about sourcing opportunities. Moreover, the strategic procurement means that the function has been given more responsibilities such as the process of planning, evaluating, implementing, and controlling highly important and routine sourcing decisions. Further, they contributed more professionally to achieve the company's long-term objectives and integrates with business practices through a rational interaction with suppliers and shifting focus from cost and value, to return on investment (ROI), and that can only be achieved through early supplier involvement, supplier development, supplier assessment, supplier certification and measurement.

## 2) The Importance of Procurement from MM Aspects

The procurement function plays the key part in Material Management and it has become more important in recent years as the success of any business in today's market is more dependent on the products being offered to its customers, and that requires to have good procurement practices in order to ensure having the right raw materials that bring benefits in terms of saving money or improving quality of the goods, works or services procured, thus, procurement is the first cost-control step of overall supply chain planning, and is worthy of serious consideration. Also, increased global scope of operations has resulted to increase the usage of outsourcing, hence the buyers' increased dependence on suppliers' capabilities will make procurement work a vital undertaking for corporations to master. The following are the key elements that make the procurement function is important within the Material Management context in an organisations:

a) **Predictability and Reliability**

The procurement function has to bring reliability to a company's business. The main role of procurement is to ensure reliability in terms of getting a product delivered to a specific point by a specific time or having a service performed whenever needed and as needed, and with such reliability the business thus can become predictable. To this end, the business can only stay predictable based on procurement's performance and thus meet the customers 'needs from the aspects of the right quality and the right quantity being delivered at the right time every time. With the lack of predictability, the business will ultimately lead to costs multiplying as unpredictability and uncertainty are the major contributors to rising costs.

b) **Globalization Aspect**

Most of today's supply chains are more globally-focused, technology-infused, interactive and collaborative, and procurement is the main source dealing externally with suppliers around the world. Therefore, procurement plays the key role to improve the business image and to ensure that products and services could smoothly reach more people. For that reason, organisations nowadays consider the function as a value-added element to the business because their decisions could either improve the business or vice versa.

c) **Profitability Aspect**

Procurement can make the business more profitable through looking at ways to save money in a number of areas within the procurement function, as they responsible for almost 60% of the turnover in a business, according to recent studies. The new trends have pushed procurement to drive down procurement costs, improve supplier terms and decrease materials prices. Three ways of making profits are:

i) **Reviewing Supplier's Terms and Discounts:** Through professionally discussions and negotiations with suppliers as to obtain procurement savings by altering the procurement patterns. For example, one of the tactics in this respect is to procure slightly more materials to receive a higher discount.

Also, letting a supplier be informed about alternatives with better options in the market in order to obtain more discounts as possible, or to give a supplier an option for a long-term contract in case getting a better price than other suppliers in the market.

ii) **Consolidating Suppliers and Deliveries:** In this way, a good buyer can include delivery charges to the total costs of supplies. Thus, processing the purchasing documentation and payment processing charges will automatically fall. That can be achieved through skilled buyers, especially through a negotiation session with a supplier.

iii) **Consolidating Purchasing Requests and Intervals:** This cuts down on delivery costs and purchasing documentation.

d) **Reviewing Purchasing Requirements**

Through a good communication with a warehouse along with updated records, procurement can be well informed about purchasing requirements, and hence it can be ensured that only strictly necessary purchases are made. It will cut down on excess costs and storage costs and is a good way to ensure that a company makes procurement savings.

e) **Purchasing from Agreed Catalogues**

Ensuring that only one brand or type of a product is purchased. Duplication can be expensive and is unnecessary. Higher orders from one supplier lead to better discounts.

f) **Having Updated Stock Level Records**

Stock left in warehouses is "dead money". It costs money to store, can deteriorate and become obsolete. Thus, procurement has to review stock levels regularly and ask for updating accordingly.

In some organisations, procurement handles the MRP process to ensure accuracy and reliability of stock levels.

g) **Reviewing the Specification of Purchased Products**

Some time lower price doesn't mean the materials can't fit the purpose, and this must be seriously considered by buyers prior to proceeding any purchase orders.

h) **Review Replacement Strategies**

Renew items only when necessary and not as a routine replacement, since there might be considerable costs of getting a replacement. So it is necessary to replace an important machinery part on a regular basis, but it is not necessary to replace most lights before they fail.

i) **Professionalism**

It is important to ensure trained buyers handle the purchasing process who are able to make cost effective purchasing and save money whenever possible. In a report published by Deloitte in 2018, stated that 51% of procurement leaders said their current teams were lacking in sufficient levels of skills and capabilities to deliver on their procurement strategy, which caused to poor performance in the procurement function.

j) **Centralization and Computerization**

By centralizing the procurement department, a company can make a huge save in staff, processes and technology. Also, the inventory and accounting systems must be integrated to avoid error and mistakes in stock levels, prices, and budget allowed for purchases.

## 3) The Procurement Objectives

In a new strategic procurement, the objectives of the function are aligned with the objectives of business strategies. At the strategic level, purchasing decisions affect profitability and business growth, and hence purchasing decisions must be in line with strategic objectives. The following are the key objectives that procurement must follow to ensure meeting the strategic objectives of an organisation:

a) **Support organisational objectives**

   The following some key objectives in this aspect:

   i) Monitor supply markets and trends, i.e. material price increases, shortages, changes in supply and interpret the impact these trends will have on the organisation.
   ii) Identify the critical materials and services required to support the company strategies in during new project development.
   iii) Support the organisation's need for a diverse and globally competitive supply base.
   iv) Develop supply options and contingency plans that support company plans.
   v) Consider quality of materials.
   vi) Continued improved communication with all stakeholders.
   vii) Develop KPIs for procurement personnel which will support the objectives and organisational goals.

b) **Buying what is value for money**

   Procurement is responsible to buy items that have high value for the company and in line with the strategic business goals. If a company wants to grow the business by offering low-cost goods, a matching procurement management objective is to negotiate low supplier prices. If a company wants to increase profitability by charging premium prices for the highest quality, the procurement in charge have to ensure that suppliers deliver the best products available. Procurement needs to review the objectives and align the value they provide with company strategies.

c) **Building a Long-Term relationship with suppliers**

This is one of the key objectives of procurement to have a strong and exclusive relationship with suppliers, in order to obtain lower prices, more reliable service and improved support. Such a business strategy may be more effective if supported by such relationships, and procurement objectives should include the pursuit of long-term relationships if they might be a strategic asset. In case the strategy is to deny the competition access to a supplier, the procurement management may have to negotiate an exclusive supply agreement. Other related objectives in this aspect are:

i) Manage the supply base, identification and mitigation of risks.
ii) Identify new potential suppliers and develop relationships.
iii) Improvement and development of non-competitive existing suppliers.
iv) Determine the method of awarding contracts.

d) **Cost Leadership**

One of the important objectives of procurement is cost leadership. A strong stance on cost leadership can help to drive significant improvement to the bottom line. This is a key when procurement is expected to demonstrate its ability to drive meaningful savings. However, the cost leadership must not negatively impact the operational efficiency, but obtaining better use of available resources and ultimately saving money.

e) **Continuous supplier evaluation**

Procurement should have a system to evaluate supplier performance as it is considered an important objective to meet the business strategy of a company, and it should be performed on a continuous basis in accordance with procurement management objectives. When supplier efficiency and performance increases in line with other business functions, then the costs of processing and production get reduced. Put in place procurement management objectives to include benchmarks for suppliers in terms of product failure rates, on-time delivery percentages, and competitiveness. Continuous evaluation against such benchmarks lets procurement identifies preferred suppliers and those with exceptional performance.

f) **Information Technology system integration with suppliers**

Integrating systems with suppliers can achieve substantial economies among partners. For example, through a sharing system a supplier is able to know the stock records and then they can directly and automatically ship products when stock runs low. Strategically plan for procurement management to support this direction and consider such integration possibilities when selecting suppliers. Specific procurement objectives might be to automate supply of products the business needs regularly; automate receiving, invoicing and payments; and integrate the system's tracking of quality issues and customer support with key suppliers. Also, from procurement side, the following procedures must be conducted to ensure accuracy of stock records:

i) Monitor spends on stock, direct charge and service spend, monthly or quarterly, and that should be done in close liaison with the warehouse.

ii) Major reagents and consumables over available stock using any analytical system, such as 80/20 Pareto principle.
iii) Review physical stock counts reported to ensure accuracy of stock on hand.
iv) Make sure if any changes made in stock categories, i.e. ABC based on the latest stock consumptions and movement.
v) Make sure MRP is run efficiently either by a warehouse or production.

g) **Support operational requirements**

These are the basic objective in procurement. The procurement function is responsible to buy products and services at the right price, the right source, the right specification, the right quantity, the right time, and to the right customer. However, the following factors as objectives must be considered and followed before placing any order:

i) Review the requirements for the material or service being provided.
ii) If possible, suggest alternative standardised materials that can save the organisation money periodic review of categories can allow greater leveraging of requirements.
iii) All purchases must go through the approved procurement processes.
iv) Engineering and other functional inputs are part of this process.
v) Contractual agreements should be done with the involvement of Procurement.
vi) Increased use of sourcing teams.

h) **Develop strong relationships with stakeholders within the organization**

Procurement should have a smooth communication with internal customers in order to achieve greater understanding of requirements and integration. The communication should be implemented based on a timely manner, pro-active approach, and sharing accurate and updated information. e.g. market trends, stock level, price, quality, delivery, forecast and demand. The most important communication must be built among warehouse, procurement, finance, and sales. Additionally, the following objective must be set up in procurement to ensure efficient operations:

i) Management of procurement staff.
ii) Development and maintenance of policies and processes.
iii) Introducing and leveraging appropriate technology that defining procurement strategy and structure.
iv) Provide procurement leadership to the organisation.
v) Provide professional training and growth opportunities for employees.

## 4) Procurement within the Organizational Structure

The structure of a procurement organisation ranges from a single person with responsibility for purchasing to a large centralized department or decentralised organisation with procurement professionals working in separate locations or business units. To this end, it is important for an organisation to set up the right structure as procurement typically accounts for more than half of an organisation's expenditure, according to latest references for Business (e.g. Dilliott, Mackansy). It also plays an important part in an organisation's competitive strategy. In a small company, for instance, the procurement function is normally handled by a small individual team in a section under the finance department to deal with purchasing transactions for other departments. It could be also dedicated individual members in each department such as production, marketing, sales, and others to purchase required products or services to meet their own departmental requirements. Therefore, the company would not have a consistent purchasing procedure as the function is considered a non-value added to the business due to a fragmented group of teams work under the allocated budget for each department.

a) **Procurement Department**

As an organisation grows, procurement then takes a different shape within the organisational structure. The procurement function is established as a separate department by having a procurement manager with professional experience and qualifications. The organisation also recruits a few key employees and a shared or dedicated administrative assistant to handle the key activities in the procurement department based on defined responsibilities, and the number of assistants may increase based on the scale of purchasing grows. In such an organisational structure, the procurement department works with a high degree of centralisation that information tends to flow in a more top-down fashion, from the business owner to the procurement manager to department employees. The department takes the responsibilities for purchasing supplies for all departments, discussing their requirements, identifying suppliers and processing orders. By coordinating procurement, the company can place larger orders with preferred suppliers. It may be able to negotiate lower rates and impose consistent quality standards on suppliers.

b) **Centralised Procurement vs. Decentralised Procurement Structure**

When a business grows, an organisation may decide whether to operate centralised or decentralised structures. In the centralised model, a single procurement department takes responsibility for purchasing on behalf of the company. The department, which may consist of a purchasing director, managers and assistants, imposes standard policies and procedures across the organisation with the aim of reducing costs, increasing purchasing efficiency and achieving consistent quality. To improve the service to different locations, the department may appoint specialists responsible for purchasing specific categories of supply. The scenario with decentralised structure is different, as the organisation must delegate procurement authority to locations and division based on the model. However, there are pros and cons of each model and thus an organisation must consider all factors and a deep dive into the pros and cons of each model. The following are the pros and cons of each model:

a) **Centralised Model**

**Pros:**

i) Better management control over money and data.
ii) Better control over suppliers records and orders being made.
iii) Approvals process more efficient and maverick spending is eliminated.
iv) Better way to negotiate deals with suppliers based on the knowledge and skills of buyers in one area.
v) The information stored in a centralized database and thus, procurement is better able to improve supplier risk management, ensure supplier diversity and comply with corporate social responsibility initiatives.
vi) Less total spending costs.
vii) Employees feel more motivated to work together in one area.

**Cons:**

i) More layers of bureaucracy to every transaction, and removes decision autonomy from local managers, which can result in job dissatisfaction.

ii) Could cause lagging processing and delivery times when the procurement department makes strategic buying decisions over local suppliers, or when an emergency situation arises in which a department needs immediate supplies.
iii) Tough competition among staff to get promoted.

b) **Decentralised Model**

**Pros:**

i) Supplies can be obtained immediately as buyers are closer to the supplier area.
ii) Local managers are in the best position to understand the needs of their divisions.
iii) Order processing is fast and easy, with no wait for approval. If a need arises, it can be sourced and filled immediately, Replacements for defective or damaged shipments can be initiated immediately, without routing through the company's procurement process.
iv) Better chance to staff to get promoted.
v) Easier to approach a supplier for negotiations.

**Cons:**

i) Opportunities for bulk purchasing across departments, and for negotiating better terms based on bulk could be lost.
ii) Orders are typically made by administrative staff, and not purchasing experts who have the knowledge and skill to evaluate suppliers, consolidate orders, and negotiate better deals.
iii) A decentralised system often means disorganized data. Compliance issues may arise as managers order or reorder for the greatest expediency as opposed to strict protocol adherence.
iv) Performing a spend analysis is far more difficult with a decentralised system, because data in different systems are rarely standardised.

v) More difficult to build a long-term relationship with suppliers compared to centralised model.

Therefore, building an organisational structure and processes of procurement need to take all factors and elements into consideration in order to develop the capabilities necessary to achieve the benefits expected by its clients, to develop and improve engagement with business partners, to participate in strategy definition and deployment, and to achieve better supplier management. This will help procurement on its way to contributing to the overall objectives and benefits of the organisation. However, the most common structure used by organisations is a functional structure which is considered the best way to achieve greater operational efficiencies, as all procurement staff with shared skills and knowledge work together and perform special tasks, as illustrated in *Figure 5.1*. The advantages of this type are:

i) *Specialisation*: The team focuses on one area of work.
ii) *Productivity*: Specialism means that employees are skilled in the tasks they do.
iii) *Accountability*: There are clear lines of management.
iv) *Clarity*: Employees understand their own and others' roles.

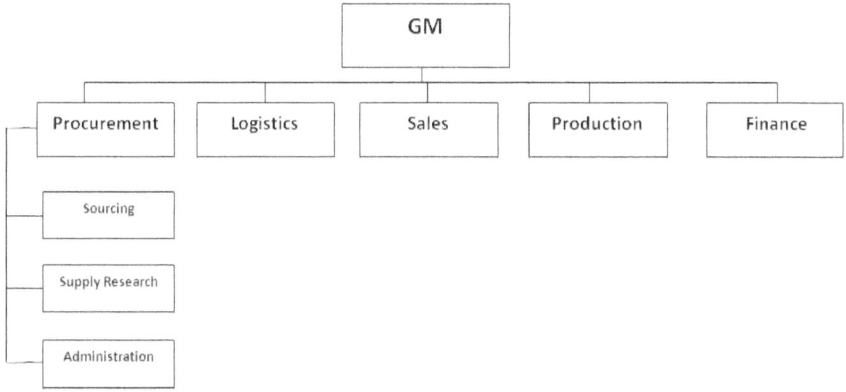

Figure 5-1 Common Organisational Structure of Procurement

## 5) Main duties of Procurement

The main duties of the procurement department are the acquisition of supplies, services, and construction in support of the business. The department is the entity within an organisation authorised to issue invitations to bid, requests for proposal, requests for quotation and issue contracts or/and place purchase orders. Moreover, the department issues purchase orders, develops term contracts, and acquires supplies and services. They also responsible to dispose all surplus property and equipment and offer the disposed items and assets for sales (Finance department takes such responsibility in some organisations).

The procurement department is very conscious of its responsibility and accountability in the expenditure of the approved budget, and should be managed in line with the guidance. Therefore, the department should maintain a competitive bidding process in accordance with general policy and regulations approved by the top management. To this point, the procurement must ethically perform duties either with internal or external customers. For example, the procurement manager must consider all reliable suppliers in the markets to participate in the various bidding processes offered, but awards shall only be made with the vendors who can prove that they can perform successfully under the terms and conditions of the proposed purchase and provide the most advantageous bid/quote according to price, quality and other requirements specified in the bid document. Consideration must be given to such things as vendor's integrity, past record of performance, and financial and technical resources. The following are the key roles and duties of procurement must be seriously considered in the job description of buyers by an organisation:

a) **Negotiating**

One of the key role and duties of procurement is Negotiating. The negotiating process is considered a large part of the procurement role, and it can also be the most difficult. In simple words, Negotiating is the process that procurement professionals go through to create favourable terms as part of a new supplier contract. This can involve negotiating different terms with an existing supplier when a contract is renewed, or discussing terms from scratch with a brand new vendor. Moreover, Negotiations are typically used to determine the fairest price and payment terms, delivery and production time, quality standards and more. The negotiations need to consider the best option for both supplier and buyer, rather than just aiming to get the cheapest possible price, as this will help to build stronger relationships with long term suppliers. This role is always seen tough and critical because purchasing negotiates is done on behalf of the company, and thus good arrangements are a must in this process, and gathering relevant information and setting objectives prior to entering into negotiations.are the most important elements in such arrangements to enure achieving a Win-Win approach.

b) **Maintaining Documents and Records**

Other important duties are to maintain all documents that pass between the procurement department and suppliers. Such job is normally handled by Administration within the procurement department, and could be a dedicated employee whose job only controls documents. All documents must be maintained safely and securely, and must be accessible by authorised staff only. Furthermore, all documents must be developed and maintained in a a central filing system to ensure their movement properly. Keeping documents and records is very important to protect the business, and it is done either physically and/or electronically. However, currently, most businesses use a document management system to improve record keeping and comply with business obligations, and such software can ensure that operations are effective and efficient. Also, it can make the job easier to capture information, generate reports, tracking records, and updating records.

c) **Effective Communication**

One of the important duties of procurement is to have smooth communications with other stakeholders within an organisation. We have discussed in the previous section that procurement has moved on from being a discrete activity to being the hub of a large part of the organisation's business activity. By its very nature, procurement must establish a strength relationship with internal stakeholders in order to solve common problems, especially with the key functions such as Finance, Warehouse, Production, Sales, Engineering, and marketing. For example, the relationship between Procurement and Engineering in terms of technical specifications of products needed for an organisation, and types of materials go into them must be procured by procurement. The relationship with Finance mainly concentrates on budget allocation and financial planning, as it forms the major causes of business success or failure. The relationship with Production is based on the material requisitions needed for manufacturing that pass to the procurement function, and with good communication and coordination between the two function many issues could pay off, such as procuring unnecessary materials, wrong materials, or costly materials, along with the time materials must be available. The relationship and coordination with Sales is based on the forecast aspect, which is important and their interest in the delicate area of reciprocity such as quality and delivery. The relationship and coordination with the warehouse function is very crucial in terms of having agreements over accuracy of stock records and stock levels available in the warehouse.

d) **Other important duties of Procurement**

   i) To procure required services, materials, equipment, and construction while ensuring that quality, safety, and cost-effectiveness are achieved.
   ii) To ensure that procurement transactions are conducted in a manner providing full and open competition whenever practicable.
   iii) To comply with procurement rules and policy in an organisation as required.
   iv) To solicit the participation of all qualified and responsible contractors and suppliers in the procurement process.
   v) To assure equity for all parties involved in the procurement process.
   vi) To obtain the best value for the money spent.
   vii) To eliminate the possibility of corruption or unethical practices in the procurement process.

## 6) Important Aspects of Procurement

Procurement is about people, i.e. buyers and suppliers, which means that transparency must be imposed by an organisation in any deals take place between them. One of the key issues in procurement is Ethics, which play a major role in procurement, especially after emerging new technologies. Being ethical means being in accordance with the rules or standards for right conduct or practice, especially the standards of a profession. Thus, the message has to be clear to buyers and suppliers that there is a responsibility to act ethically, because failing to abide to ethical practices can lead to immoral and illegal practices such as bribery, favoritism, illegal sourcing etc. It is important to note that procurement ethics, like many other aspects of management, are top-down. The example set by senior management, its attitude, and its behaviour, strongly influence employees at other levels. Senior managers have to demonstrate fairness and transparency to encourage the visibility of the same qualities in procurement executives and teams. Ethical procurement has to be on display.

a) **The importance of ethics in Procurement**

Ethics is the tone for how the entire company must be run on a day-to-day basis. Having a foundation of ethical behavior is the key factor to create long-lasting positive effects for an organisation, including the ability to attract and retain highly talented individuals, and building and maintaining a positive reputation within the community. In some organisations, there is a an ethics ombudsman whose main job is to look after all ethical issues in the whole business, and with whom an employee can confidentially communicate any real or perceived ethical violations. Further, the ethical philosophy is able to build a stronger bond between individuals on the management team and create more stability within the organisation, and employees are able to make better decisions in less time with the existing business ethics in terms of productivity and morality. For example, when employees complete work in a way that is based on honesty and integrity, the whole organisation will benefit by increasing the performance of staff at a higher level with more loyalty to the organisation. Also, the ethical operation is directly related to profitability and reputation of an organisation, based on the trust it could build to motivate more investors to deal with the organisation, especially in procurement whose main tasks to deal with external clients.

b) **The advantages of ethical standard in Procurement**

It is important to uphold an ethical standard in procurement to achieve the following goals:

i) To avoid acting any unethical behaviour in doing business with clients as they usually have control over large sums of money.
ii) Ethical behavior in conducting business helps in establishing a long term relationship and goodwill with suppliers.
iii) An ethical person is respected in the business community. Once a buyer earns a reputation within an industry, it is difficult to change it. A professional reputation is something a buyer carries throughout his entire career.
iv) To ensure practice integrity by all employees in the procurement department.
v) To avoid conflicts of interest and personal enrichment through treatments all suppliers fairly and equally.
vi) To ensure sustainable and reliable relationship with suppliers in a long term perspective.
vii) To ensure employees in procurement comply with legal and other obligations.
viii) To avoid any kind of corruption that could occur in cahoots in the procurement department.

c) **Lack of Ethical standard in Procurement**

The following immoral or illegal practices could occur in case the absence or lack of clarity in ethical standards in procurement:

i) ***Bribery:*** Payments in cash or in kind made to individuals in procurement to buy materials from a supplier. Bribes can occur before, during, or after (kickbacks) award of a contract.

ii) ***Coercion:*** Threats made against or pressure put on individuals with the same objective as bribery to gain support for a supplier or contract negotiation. The difference is that whereas bribery aims to motivate individuals with what they can gain, coercion aims to motivate through the fear of what they might suffer or lose.

iii) ***Extortion:*** Asking for a bribe or similar illicit payment. This may or may not be accompanied by a menace.

iv) ***Favouritism:*** Also known as nepotism, in which individuals give undue preference or negotiating advantage to a supplier who is a friend or part of the same family.

v) ***Illegal sourcing****:* Suppliers offer misrepresented goods or services, or produced illegally or immorally, whether because of the materials already used or the labour conditions in which production takes place. Stolen and black-market supplies are good examples.

vi) ***Traffic of influence:*** The exchange of an award of contract for a favour or preferential treatment by the other party of another individual or organisation.

# CHAPTER 6

# Distribution

## 1) Introduction

Distribution is the third critical function in materials management. The distribution management is about to overseeing the movement of goods from a supplier or manufacturer to the point of sale, and it includes many activities such as packaging, inventory, warehousing, supply chain and logistics. We already covered some activities in the previous chapters, but in this chapter the concentration is only on the activities related directly to the distribution aspect in order to cover all activities of Materials Management before covering the technology applications in all three components of materials management in part 3.

Companies pay high attention to this part as it plays a key role in fulfilling customer demands, ordering and managing inventory, controlling inbound and outbound shipments, reducing costs, saving time, and meeting company objectives. Furthermore, the distribution brings processes and discipline to transportation management through effective utilising a database of partner carriers, and build the most optimal transportation approach for the business. Also, it is the key factor in handling outbound, inbound, and internal transportation concerning inventory control, packaging, handling customer relations, order processing, and marketing. In essence, the distribution involves controlling the movement of materials and goods from their source to their destination. Many organisations look at the distribution from the marketing perspective as its role is mainly associated with a major component of the marketing mix, and Marketing always determines their decisions based on the performance of physical distribution, such as (i) tracking system (ii) distribution channels and transportation methods used (iii) the best way information could be shared.

## 2) Importance of the Distribution function within the MM Context

The distribution role within the material management context is very important from different aspects. It provides a new orientation for marketing and sales, as it confers place-utility and time-utility to a product by making it available to the user at the right place and at the right time in that way to maximises the chance to sell the product and strengthen the company's competitive position. Without the distribution function, a company is not able to transport products to a place of consumption, along with storing products for a considerable length of time if required. The following factors made the distribution role becomes crucial within the MM context:

i) **Creating Time and Place Utility:** The distribution activities help in creating time and place utility. This is done through transportation and storing. Transportation system creates place utility as it makes available the goods at the right place where they are required, and warehousing creates time utility by storing the goods and releasing them when they are required. The function also provides the place and time dimensions which contribute a basic element of the marketing mix.

ii) **Cut in Distribution Costs:** The distribution cost accounts for a major part of the price of the product, and the costs estimated in the range of 20 to 25 percent of the total price. Therefore, If the costs of the distribution are handled properly, then it can be instrumental in slashing down such costs. There are different ways to handle the distribution to control related costs, which could be either tangible or intangible, such as proper and systematic planning of transportation schedules and routes, warehousing location and operation, material handling, order processing, communication, training, etc.

iii) **Improved Consumer Services:** The distribution function plays an important role to ensure products in right quantity available at the right time and right place (consumer's location), and that are considered the key factor to keep consumers continue dealing with a company. The role of the distribution extends to cover more services in this aspect, such as having an effective communication with customers throughout the distribution process, after-sale services, product range offering, product availability and the like. Moreover, the distribution's role can determine a service level through the measuring of how well the customer service function is being performed.

iv) **Helps in the Stabilization of Price:** The distribution role is able to make the stability of prices through proper use of transportation and warehousing facilities that can bring about amicable and matching adjustment between the demand for and supply of goods thus preventing price fluctuations and distortions.

v) **Helps to increase Market Share:** An efficient distribution system can contribute towards increasing market share as it is the major aims of any growth aspiring business in today's competitive market. A company can increase its market share through a well designed distribution function, e.g. decentralising the warehousing operations; devise the combinations of efficient and economic means of transport to pen errata into the areas untapped so far, planning inventory operations to avoid stockouts and gluts. In other words, reaching customers and meeting their demands wherever they are to gain their loyalty.

## 3) Objectives of the Distribution Function

The main objectives of the distribution function are to decrease costs and transit time while increasing customer service to obtain a fully customer satisfaction. In other words, the distribution management must strive to make balance service, distribution costs, and resources. In the current competitive business, the distribution function is considered a key factor to make the business distinguished through adding more value to customers than competitors. The following are the main objectives of the distribution function that should be set up by organizations:

i) **Achieving Consumer Convenience:** Of course, this is the main objective of the distribution function, through using the right kind of distribution method to increase consumer convenience. Further, a company must think rationally how to let consumers choose its products by making their needs attended at any time from the convenient place, and at better price than competitors. Also, whether using own services or through middlemen, the company must offer consumers a chance to select the most suitable products that make them fully satisfied with the services being offered.

ii) **Facilitating Continuous Production:** The distribution function is important in continuous production to meet consumers' demand. An efficient distribution network facilitates continuous production because of the sophisticated storing facility, suitable and rapid means of transportation, efficient communication channel, access to local and global market, advance ordering, buying incentives to sell in the off-seasons, rapid ordering and executing, etc.

iii) **Achieving Economy:** This objective can't be achieved without a suitable and efficient distribution system. Economising the distribution role through an efficient system will result in reducing overall costs in a number of ways. For example, speedy order processing, availability of the latest transportation and communication, benefits of scale of economy, rapid sales turnover, insuring the products, and many other similar benefits lead to low costs, and ultimately low selling price.

iv) **Reducing Degree of Damage & Wastage:** The distribution management can reduce product damage that could happen during storage, transportation, and handling. Beside, availability of insurance at a lower premium can reduce considerable risk during storage and transportation. Further, having cold storage, rapid and safe means of transportation, and other facilities relating to distribution can reduce the damage or wastage of product. Reduced damage and better quality significantly contribute to the success of the product.

v) **Increasing Competitiveness:** The distribution management can strength business' competitive by establishing a systematic distribution network. Many companies can distinguish their offers by availing products differently than competitors. An effective distribution system will result positively to services, availability, timing, price, and similar benefits. Undoubtedly, if all the components of distribution work effectively, the distribution role can be a powerful means to fight with competitors.

vi) **Lowering Idle Stocks:** This objective relates how to control stock efficiently. Producers and distributors can minimise reordering size or safety margin by the effective distribution system. Due to speed and precision in placing and executing orders, and advanced ordering by distributors, they are not required to maintain more stock of finished products. This facility can reduce overall inventory costs and need of working capital.

vii) **Ensuring Continuous Availability:** This objective concern by offering a direct benefit to consumers. Due to wide availability of products, consumers are not required to store the essential commodities. They can buy the right quantity as and when they need. It leads to several benefits to consumers.

viii) **Achieving Rapid Turnover of Stock:** The distribution role is also targeted to speed up the turnover of stocks. From an investment of cash in raw materials to realisation of cash through the sales of finished goods can be speeded up. Stocks can be speedily converted into cash. So, the duration of the working capital cycle can be reduced, and the need of working can be minimised.

## 4) Elements of the Distribution Function

The main elements of the distribution system in some details are:

a) **Customer Service**

Customer service is a predefined standard of customer satisfaction, which a company plans to provide to its customers. The distribution responsibilities in this aspect are:

i) Ensuring the product availability all the time.
ii) Improving order cycle time between placing the order and its delivery time.
iii) Quick deliveries in case of urgent orders.
iv) Provides training to sales persons and employees engaged in transportation.
v) Having a contingency plan in case of natural or unforeseen problems, and/or loss in transit.

b) **Order Processing**

Order processing, alternatively known as order fulfillment is the handling of customer orders within the distribution center, involving the keying of customer and order details into the computer system in order to produce the invoices for the picking. The basic idea is to deliver the orders as per customers' wants of place and timing. Further, order processing is considered as the key to customer service and satisfaction. It includes receiving, recording, filing, and assembling of products for dispatch. More importantly, the amount of time required from the dates of receipt of an order up to the date of dispatch of goods must be reasonable and as short as possible. The distribution management must ensure there is no gaps exist before delivering products to consumers in order to ensure high customer service levels.

c) **Inventory Control**

Inventory control is a major component of an organisation's distribution system. It includes money invested in inventory, wear and tear and possible obsolescence of the goods with the passage of time. Therefore, inventories must only be held to meet market demands promptly, and the distribution management is responsible to ensure the inventories are delivered to consumers. The basic roles of distribution are to ensure the following:

i) Right quantity to deliver.
ii) Inventories delivered at the right time.
iii) Right Quality.
iv) Right mode of transportation.
v) Right communication means to be used with stakeholders e.g. Marketing, Warehouse, Sales, and Customers.
vi) Maintain documents.

d) **Storing Finished Goods**

Warehousing involves storing goods until they are transported to the customer upon receipt of order. This function basically involves receiving the merchandise, breaking bulk, storing and loading for delivery to customers as per their details. Warehouses usually keep goods for long periods while distribution centers operate as central/middle locations for quick movements of goods to retail stores. The distribution management must build intimacy with the warehouse through an effective communication and coordination to ensure providing the best service to customers. Today, many organisations have regional distribution centers where shipments are broken down to small loads that are then quickly transported to retail outlets as per the outlet's requirements.

e) **Transportation Mode**

Transportation is indispensable for physical distribution of goods and services, and it is considered the key element to ensure goods are delivered faster with secure transporting and handling of goods during the transit process. The distribution's role is to choose the best mode to transport goods to consumers, and that will ensure channel members like producers, wholesalers and retailers to make goods and services available at the customers' place of purchase or at their doorstep, and that must be coordinated with the marketing department. From Marketing' side, they have to decide to (i) what mode or combination of modes of transportation should be used to transport products to warehouses and from there to customers (ii) Should the transportation cost be reduced and the desired levels of customer service still maintained. Some modes of transportation are: i) Rail, ii) Airway, iii) Roadways, iv) Waterways, and v) Pipelines.

The following are some important aspects must be taken into account when selecting a mode of transportation:

i) Can be used to transport goods in proper quality and quantity?.
ii) Can be used to transport goods at the right time and at the right place?.
iii) Can satisfy customers' demands?.
iv) Cost considerations?.
v) Drivers are trained and authorized to use it?.
vi) In case of a middleman, ensure they have all capabilities to handle it?.

f) **Materials Handling**

Materials handling involves moving products/raw materials in and out of a stock, i.e. movements from a common warehouse to various storing locations. From the distribution aspect, handling services and protective packaging are the most important part to improve the level of customer service and at the same time lowered physical distribution costs. Also, material handling and packaging services have also speeded up the order processing and movement of consignments. Some products or materials may require proper handling and the utmost care and thus, a suitable handling equipment and mode must be used to ensure safe packaging and handling. The following type of handling equipment to be used depends upon the following reasons:

i) Mode of transport: rail, air, water, and road.
ii) Nature and size of goods (e.g. Materials: heavy, light, solid, liquid or gases).
iii) Place of operation: warehouse & selling floor.

In order to perform the mentioned elements effectively, the distribution management must coordinate all activities into an effective system to provide the desired customer service in the most efficient manner. Also, the distribution manager must have an effective communication with the subordinates, and guide and motivate them to make continuous improvements.

## 5) Channels of Physical Distribution

The distribution channels (also called a marketing channel as it represents one of the classic 7Ps, product, promotion, price, placement, people, process, physical evidence) refers to a chain of businesses or intermediaries through which a good or service passes until it reaches the end consumer. The channel or the route can be as short as a direct interaction between the company and the customer or can include several. It includes wholesalers, retailers, distributors, and the internet has been added after emerging new technologies. The channels are normally broken into direct and indirect forms: A direct channel allows the consumer to buy goods from the manufacturer, and an indirect channel allows the consumer to buy goods from a wholesaler or retailer. Therefore, the distribution channel can be defined as a set of interdependent intermediaries that help make a product available when, where, and in which quantities the end customer wants. The distribution channels can be very simple, with just two layers (producer and consumer), or very complicated, with several levels, as is explained below:

a) **Types of Distribution Channels**

   i) **Channel 1:** The first channel is the longest because it includes all four: producer, wholesaler, retailer and consumer. This type contains two intermediary levels; a wholesaler and a retailer. A wholesaler typically buys and stores large quantities of several producers' goods and then breaks into the bulk deliveries to supply retailers with smaller quantities. For small retailers with limited order quantities, the use of wholesalers makes economic sense. This arrangement tends to work best where the retail channel is fragmented, i.e. not dominated by a small number of large, powerful retailers who have an incentive to cut out the wholesaler. A good example of this channel arrangement is the distribution of soft drinks and drugs.

   ii) **Channel 2:** Through this channel the producer sells directly to a retailer who sells the producer's product to the end consumer. It contains only one intermediary, i.e. retailer. The goods are sold directly to large retailers which then sold onto the final consumers. A good example is Dell, which sells its products directly to reputable retailers.

   iii) **Channel 3:** Is called a direct marketing model, where the producer sells its product directly to end consumers/customers. In this case the manufacturer sells directly to consumers/customers to make the shortest distribution channel possible.

An example of a direct marketing channel would be a factory outlet store. The Bakery industry is a good example in this regard, as they sell products directly to consumers.

b) **Flows in Physical Distribution Channels**

Flows refer to various intermediaries that make up a distribution channel to be connected, such as the following types of flow:

i) **Physical flow:** It describes the movement of goods from raw material that is processed in various stages of manufacture until it reaches the final consumer. In the case of raw materials of towel which is cotton yarn, they flow from the grower via transporters to the manufacturer's warehouses and plants.

ii) **Title flow:** Is the passage of ownership from one channel institution to another; when manufacturing towels, title to raw materials passes from the supplier to the manufacturer. Ownership of finished towels passes from the manufacturer to the wholesaler or retailer and then to the final consumer.

iii) **Information flow:** It involves the directed flow of influence from activities such as advertising, personal selling, sales promotion and publicity from one member to other members in the system. Manufacturers of towels direct promotion, and information flows to retailers or wholesalers, known as trade promotion. This type of activity may also be directed to end consumers, i.e. 'end user' promotion.

c) **Strategic Channel Choices**

Important factors in formulating a channel policy are based on the following choices by an organisation, and such policies are usually developed between Distribution, Sales, and Marketing.

i) **Intensive distribution:** Is a form of marketing strategy under which a company tries to sell its product from a small vendor to a bigger store. Virtually, a customer will be able to find the product everywhere he goes. This is most common when customers purchase goods frequently, e.g. Soft drinks and cigarettes are some of the examples on which intensive distribution is followed. The aim is to achieve maximum coverage.

ii) **Selective distribution:** Where products are placed in a more limited number of outlets in defined geographic areas. An advantage of this approach is that the producer can choose the most appropriate or best-performing outlets and focus effort on them. Selective distribution works best when consumers are prepared to "shop around". In other words, they have a preference for a particular brand or price and will search out the outlets that supply. Selective distribution seeks to show products in the most promising or profitable outlets, e.g. high-end designer clothes.

iii) **Exclusive distribution:** Where products are placed in one outlet in a specific area. This approach is an extreme form of selective distribution in which only one wholesaler, retailer or distributor is used in a specific geographical area. When the firm distributes its brand through just one or two major outlets in the market, who exclusively deal with it and not all competing brands, and the dealer is required not carrying competing lines. This is a common form of distribution products and brands that seek a high prestigious image. e.g. an exclusive franchise to sell a vehicle brand in a specific geographical area, in return for which the franchisee agrees to supply an appropriate after sales service back-up. The firm hopes to get the benefit of aggressive selling by following this approach.

## 6) Transportation

Transportation is one of the core components of the distribution system, and it has been a major contributor to the economy and a competitive force in business' today. Further, it is considered as a major influence on the customer's satisfaction and supply chain activities, because if the performance of transportation falls short of expectation in any circumstances, the customer is dissatisfied and hence causing a great loss of a business, along with jeopardising the source of procurement of materials; goods and services, movement or people and even course increase in prices and loss of lives. To this end, it is important to discuss the aspects of transportation within the distribution context. Transportation is defined as moving products from the plant or warehouse of the seller to the receiving facilities of the buyer, either locally or globally. It involves two parties, carriers and shippers. Carriers are those companies that provide transportation facilities to others, and shippers are those organisations and individuals such as manufacturers, middlemen, customers, and others to whom the carriers provide transportation services. Transportation is the key element as it affects the marketing activities, because the area of a market depends on the availability of transport facilities.

## Modes of Transport

Mode of transport simply refers to the various means by which materials can be conveyed to final customers or to their final destination. There are various modes as explained below:

a) **Road Transport**: Is the most dominant mode of transporting item from one point to another; these allow for the thorough transport of goods from factory or warehouse direct to customers premises by applying the road. The road is considered the main means of transporting goods in countries with small size like Bahrain. There are different types of vehicle and bodies used in the road transport, and all depends on the types of products being loaded. The following are some examples:

   i) **Platforms:** This type provides all round access to the load, but offers little security or protection from the weather. Loads also need to be restrained. This will generally involve roping and sheeting, which is a time consuming operation.

ii) **Van body:** The van or box body reduces the payload of the vehicle, but provides protection for a perishable product and added security. Construction will depend upon the needs for insulation, waterproofing or strength. Access is usually provided by a rear door. Sometimes a door will be built into one, or both, of the body sides.

iii) **Curtain sided bodies:** Curtain sided bodies overcome the disadvantages of access, since the curtains can be pulled back to reveal the full length of the platform. This improves the speed of loading as well as unloading. Advantages of load restraint and weather protection are maintained, while body weight is less than the box body. Other variants will replace the curtains with sliding panels.

iv) **Tankers:** Tankers are designed to carry powders or liquids. They require a pumping mechanism and piping to discharge the load.

v) **Bulk carriers:** Bulk carriers are generally built as the box bodies without the roof. They require a tipping mechanism to allow the load to be discharged.

vi) **Drawbars:** A rigid master truck with a drawbar trailer is the usual configuration. The bodies may be of the demountable type. Drawbars offer increased cubic capacity for bulky lighter loads.

An organisation can use its own vehicle or through a middleman, and each one has advantages and disadvantages. For example, if an organisation decides to acquire its own vehicles, there are a number of elements must be considered such as, the type of vehicle, in terms of the chassis-cab and the body and security issues. Running own vehicle has some advantages such as:

i) Vehicles can be built specifically to carry a particular product. Special equipment for materials handling can be attached.

ii) The driver can be specially trained and will fulfil the 'ambassador' role for the organisation.
iii) Vehicles can carry the company livery, perhaps the aid organisations logo and, where appropriate, the Red Cross management retains total control over the vehicle and its operation.
iv) May be better the cost effective aspect than third party vehicles, especially in countries that the road mode is the major means.

However, there are disadvantages as well for having own vehicles such as:

i) Requires a great deal of management time.
ii) Requires specific expertise and significant capital investment.
iii) Maintenance & insurance costs are very high.
iv) Difficulty in terms of train new drivers in case a trained driver leaves the company.

In contrast, third party carriers can be considered a better option in terms of costs and transport facilities, but careful consideration must be given to the level of service required.

Some advantages of third party carriers are:

i) Better to meet fluctuating demand requirements.
ii) Better facilities in terms of variable loads and journeys can be catered by a third party.
iii) A third party is able to offer a haulier who can offer a more cost-effective and a more efficient service.
iv) Administration of vehicles and drivers are no longer the responsibility of the organisation, and thus management can concentrate more on core business activities.

v) There is no requirement for capital to be invested in transport.

There are also some disadvantages with a third party carrier like:

i) A measure of control could be lost with third party operations.
ii) Performance feedback and communication with customers needs to remain a strong feature and be controlled by the contracting organisation.

b) **Rail Transport:** This mode is particularly used for bulky and heavy consignments that require movement over medium to long distance and where speed is not vital. Rail transport is considered a safe land transportation system when compared to other forms of transportation, and it is suitable for high levels of passenger and cargo utilisation and energy efficiency. Rail transport is less flexible and more capital-intensive than highway transportation is, when lower traffic levels are considered. From a cost perspective, rail transport costs are considered less than air or road transport. In an article published by "The Economic Times" in 2016, experts believe that the rail transport system costs are approximately six times lower than the road for comparable levels of traffic, especially in the far-east countries. However, there are some disadvantages in rail transport compared to other modes, such as:

i) It lacks the versatility and flexibility of motor carriers since it operates on fixed track facilities.
ii) It provides terminal to terminal, rather than point to point delivery services.
iii) Very slow.

In countries in the Arabian gulf, rail transport is only used in Saudia Arabia since the country is the biggest by size, other five countries (Bahrain, Qatar, Kuwait, UAE, and Oman) do not have a rail transport system because of their small size.

c) **Air Transport:** The use of air transport as an alternative transport mode has grown rapidly in recent years due to availability of more facilities and technologies. Air transport has become necessary essential in current competitive markets with spreading of technological means when the speed of delivery is an important factor. The air mode provides the delivery with speed, lower risk of damage, security, flexibility, accessibility and the best frequency for regular destinations, yet the disadvantage is high delivery fee. Also, air transport is the best option in case of flooding and conflict situations where road access is difficult. Air transport is provided through:

i) Air carriers using world airlines (e,g, TransGulf Maritime Services in Bahrain).
ii) Global logistics service providers (e.g. Global Logistical Services WLL in Bahrain).
iii) Air charters (e.g. DHL, Aramex, Fedex).

There are different factors influencing the decision taken by a company to the nature of the aircraft chartered:

i) Availability and cost.
ii) The nature, quantity, weight, size and volume of the cargo.
iii) Equipment and handling available at origin and destination.
iv) The distance to be travelled and possible constraints on certain airspace.
v) The ability of certain airports to handle particular types of aircraft.
vi) Possible noise and/or operating hours restrictions at certain airports.
vii) Securing landing and over-flight permission.
viii) Availability of customs and/or immigration at the airport.

**Important considerations of Air transport:**

i) **The air Waybill (AWB):** It is a document serves as a i) receipt of goods by an airline (carrier) and ii) as a contract of carriage between the shipper and the carrier. The document includes i) conditions of carriage that define (among other terms and conditions) the carrier's limits of liability and claims procedures, ii) a description of the goods, and iii) applicable charges. The AWB is in a standard format globally in accordance with IATA definitions, and it is a non-negotiable instrument. This means it does not specify on which flight the shipment will be sent, or when it will reach its destination. However, the document is made out to a named consignee who is authorised to receive the consignment from the carrier.

ii) **Package labelling:** This is an important consideration. Package labeling is any written, electronic, or graphic communications on the packaging or on a separate but associated label. In Air transport, limited space on aircraft will require packaging, plus cargo, to be within the allowable weights and dimension for that specific aircraft. Unit load devices vary and the specific requirements needed to be coordinated before final packaging to avoid delays. However, the method of loading and unloading and onward transit may still require a strong and durable packaging medium. Ultimately, it is the nature of the goods being transported that will determine the precise nature of the packaging.

d) **Sea Transport System:** This mode is normally used for a large shipment and large quantity of raw materials and products across seas, rivers and canals throughout the world. It is only convenient for bulky pre-planned consignments which doesn't require an immediate delivery. Further, sea transportation is conducted in containers which vary in size. Goods can be grouped into a container either in Less than Container Load (LCL) or Full Container Load (FCL), along with sea tankers are used for bulk shipments of loose goods such as oil, grain and coal. Many organisations use the sea transport mode as a clever choice due to relatively low transport and handling costs, and it is considered as relatively little pollution, especially in comparison to air freight. Like other modes of transportation, Sea transport has certain advantages and disadvantages as explained below:

Advantages:

i) Ideal for transporting heavy and bulky goods.
ii) Suitable for products with long lead times.
iii) Costs are less than other modes like Air transport.
iv) Suitable for the different types of industrial goods and consumer goods.
v) Marine and cargo insurance cover all risks for loss or damage of goods.

Disadvantages:

i) Longer lead/delivery times.
ii) Bad and sever weather could cause some delays.
iii) Difficult to monitor the exact location of goods in transit.
iv) Customs and Excise restrictions.
v) In some circumstances, it could be costly.

**Important considerations in the sea transport mode**

Bills of Lading (B/L) and Sea Waybills (SWB) are the two main documents used in the transport of goods by sea, both domestic and abroad. It can be confusing to apply and use these documents correctly; however, each one performs very specific functions. In brief, a Sea Waybill is evidence of a contract of carriage and receipt of the goods being transported; whereas a Bill of Lading acts as the contract of carriage and receipt of the goods, while also serving as a document of title affording ownership. Further, the B/L is a must in Sea transport being issued by the carrier, transport agent, shipping company, vessel operator, and even the captain of the ship, and its importance is connected to the following functions:

i) Receipt of the goods sent on a vessel and their apparent condition.
ii) Evidence of the contract of carriage.
iii) Document of title to the goods, transferring ownership to the holder, who may collect the goods from the carrier at the destination port.
iv) It is a negotiable instrument accepted by banking institutions.

Whereas the Sea Waybill (SWB) is used when the shipper decides to release ownership of the cargo immediately. This means that the goods can be delivered to the person identified in the document, and they will simply have to verify their identity instead of presenting a document to claim the freight. SWB is only considered an evidential function and does not give title to the goods (non negotiable). Also, when the shipment is loaded, the shipper receives a SWB simply as a reference. In this case, neither the shipper nor the importer is obligated to submit any additional documents to the carrier, and therefore the cargo is released as soon as it is available at the port. A SWB is suitable under the following circumstances:

i) When there is a high degree of trust between the shipper and the consignee.
ii) When the goods will not be traded or sold during transport.
iii) When the goods are paid for with an approved line of credit.

## 7) Incoterms

Freight incoterms are trade terms that are used as key elements of international contracts of sale. They tell the parties what to do with respect to the carriage of the goods from buyer to seller, and export & import clearance. They also explain the division of costs and risks between the parties. In simple words, they refer to how far along the process will the supplier ensure that the goods are moved, and at what point does the buyer take over the shipment process. The following are definitions for each incoterm rule:

a) **EXW**: It refers to Ex Works, the seller delivers when it places the goods at the disposal of the buyer at the seller's location or at another specified place. The seller does not need to load the goods on any collecting vehicle, nor does it need to clear the goods for export.
b) **FCA:** It refers to Free Carrier, the seller delivers the goods to the carrier or another person nominated by the buyer at the seller's location or another named place.
c) **FAS:** It refers to Free Alongside Ship, the seller delivers when the goods are placed alongside the vessel nominated by the buyer at the named port of shipment. The risk of loss of or damage to the goods passes when the goods are alongside the ship, and the buyer bears all costs from that moment onwards.
d) **FOB:** It refers to Free On Board, the seller delivers the goods on board the vessel nominated by the buyer at the named port of shipment. The risk of loss of or damage to the goods passes when the goods are on board the vessel, and the buyer bears all costs from that moment onwards.
e) **CPT:** It refers to Carriage Paid To, the seller delivers the goods to the carrier or another person nominated by the seller at an agreed place. The seller must pay the costs of carriage necessary to bring the goods to the destination.
f) **CFR:** It refers to Cost and Freight, the seller delivers the goods on board the vessel. The risk of loss of or damage to the goods

passes when the goods are on board the vessel. The seller must contract for and pay the costs and freight necessary to bring the goods to the named port of destination.

g) **CIF**: It refers to Cost, Insurance and Freight, the seller delivers the goods on board the vessel. The risk of loss of or damage to the goods passes when the goods are on board the vessel. The seller must contract for and pay the costs and freight necessary to bring the goods to the named port of destination. The seller also contracts for insurance cover against the buyer's risk of loss of or damage to the goods during the carriage. The seller is only required to obtain minimum insurance coverage. However, should the buyer wish to have more insurance protection, it will need either to agree with the seller or to make its own extra insurance arrangements.

h) **CIP:** It refers to Carriage and Insurance Paid To, the seller delivers the goods to the carrier or another person nominated by the seller at an agreed place. The seller must pay the costs of carriage to bring the goods to the destination. In addition, the seller is required to obtain minimum insurance coverage. Should the buyer wish to have more insurance protection, it will need either to agree with the seller or to make its own extra insurance arrangements.

i) **DAT:** It refers to Delivered At Terminal, the seller delivers when the goods are unloaded and are placed at the disposal of the buyer at a named terminal at the named port or place of destination. The seller bears all risks involved in bringing the goods to and unloading them at the terminal at the named port or place of destination.

j) **DAP:** It refers to Delivered At Place, the seller delivers when the goods are placed at the disposal of the buyer on the arriving means of transport ready for unloading at the destination. The seller bears all risks involved in bringing the goods to the named place.

k) **DDP**: It refers to Delivered Duty Paid, the seller delivers the goods when the goods are placed at the disposal of the buyer, cleared for import on arrival and ready for unloading at destination. The seller bears all costs and risks involved in bringing the goods to destination and is obligated to clear the goods not only for export but also for import, to pay any duty for both export and import and to carry out all customs formalities.

FOB (Free On Board), EXW (Ex Works) and FCA (Free Carrier) are the most Incoterm rules used by companies around the world, because they are written from a legal perspective. Most companies in Bahrain use these three rules to avoid confusing and misunderstanding in a contract signed with shipping agencies.

## 8) Packaging

Packaging plays a key role in the distribution system, and it can affect sales and consumers' purchasing decisions when intending to buy a product, and it is a prerequisite for every product because only a packaged product that is transported, stored, carried, etc., in the supply chain. Packaging from a logistics point of view is the general term applied to cover all aspects of the applications of materials and procedures in connection with them to the preparation of goods for handling, storage, marketing, and dispatch. From the environmental perspective, packaging is a necessity for the containment and protection of products, for the protection of the environment from the products. From the marketing aspect, the packaging enables more efficient distribution and storage of products, which means it can help to reduce costs and cut lead-times in the supply chain. Furthermore, packaging has also a significant impact on productivity and customer service as it is an important element in sales promotion, customer attention, and brand communication.

   a) **Functions of Packaging**

   There are three important and interrelated functions of packaging; protection, utility and communication. There is an increasing trend to view packaging in terms of the functions and value that it provides, rather than just in terms of traditional materials, as it is part of a total system, with responsibility to minimise the cost of delivery as well as to maximise sales. The following are the main functions of packaging:

i) **Protection:** It is an important packaging function because spoilage and damage while products being distributed is a waste of production and logistics resources, and add further costs along with lost customers. To this end, well designed packaging is important to protect products during various atmospheric conditions, storage conditions, stresses during warehouse and vehicle stacking in delivery trucks, and shocking from vibration during transport and handling. Also, packaging can help unitise the pallet load and provides protection to the packages against moisture, dirt and abrasion in a warehouse (indoor or store yard). Therefore, technical packaging functions are designed into packaging, so as to facilitate safe distribution of the product.

ii) **Productivity:** Productivity in logistics is a very important aspect to control labour and capital intensive. Packaging productivity has an important role in this aspect as a result of the output depends on the number of packages loaded into a truck. Therefore, there must be some initiatives by a company through unitisation and size reduction in order to increase the output of logistical activities. A good example is palletisation, which dramatically improves the productivity of most material handling operations compared to break-bulk handling. Unitisation enables a single person and a forklift to handle thousands of kilograms in an hour.

iii) **Packaging Utility:** Packaging plays an important role in protecting workers in logistics activities. Most injuries in physical distribution activities involve shipping containers.

There are two types: accidents, usually involving an unstable package falling on a person, and chronic stress injuries due to manual handling of goods. Routine manual handling of packages has always been taken for granted, but it has a reputation for causing chronic back injuries. Many warehouse workers are hurt by the packages that are heavy, bulky, or must be lifted to a top shelf.

iv) **Communication:** One of the core functions of packaging is to use it as a means of communication with customers through product information, and creates feelings and associations at a psychological level by branding and positioning the product in the mind of the consumer. The communication through packaging is direct with text describing product attributes or subtle by using colors, size, images, graphics, material, smell, name of the product, brand, country, information about a product, special offers, instructions on usage, simplicity and ecology of the package, ergonomics, and innovativeness. From marketing's point of view, communication plays an important role to let consumers be familiar with a product and hence, it increases his/her perception of the product's value and relative status on the ladder of its product category, and thus motivates them to purchase the product. On the other hand, communication is important for inventory records in a warehouse by having a correct identification of stock-keeping units (SKU) such as SKU number, name, brand, size, colour, lot, code dates, weight and number in the package are critical for good information management, and that will help to have accurate information about each stock item, and workers will be able to quickly recognise a package from its label. Finally, packages can communicate how to use, store, recycle, or dispose of the package or product in both indoor and outdoor warehouses.

v) **Security:** packages can play an important role in reducing the risks associated with the shipment. In a smart warehouse, electronic devices like RFID tags are installed on packages to identify the products in real time, reducing the risk of thefts, and increasing security.

b) **The Packaging System**

Packaging is built up as a system usually consisting of a primary, secondary, and tertiary level.

i) **The primary package:** is the packaging that actually houses and most protects a product. Such package is retained till the consumer is ready to use the product. For example, plastic packet for socks while in some other cases such package is used throughout the life of the product such as the bottle carrying jam or tomato sauce etc. In simple words, it is the packaging the consumer takes home.

ii) **The secondary package:** It relates to the issues of visual communication and it is used to group primary packages together. Such packing is retained till the consumer wants to start using the product. For example. Pears Soap usually comes in a cardboard box. Consumer first throws the box when he/she desires to use it and then discards plastic wrapper too to get hold of the soap. Simply, it is the packaging around packaging used to group items together such as boxes, trays, film wrap; in other words, the packaging around your packaging.

iii) **The tertiary package:** Is used for warehouse storage and transport shipping. Such packaging is mainly used for transporting and warehousing, such as cartons, corrugated boxes, wooden crates, and pallets. The tertiary package is very important for the purposes of protection, transportation, and handling.

## 9) The Distribution Function in Organizational Structure

The distribution function is assigned either to Logisitcs or Sales within the organisational structure chart, as explained below:

a) **Within Sales**

   The distribution function, from a strategic perspective, is always considered as the main tool for both Sales and Marketing to cover up all customers' needs in maximum areas possible, and that is the purpose of the function in all companies in current competitive markets. Further, the physical distribution function with its logistical resources is assigned as part of the sales function to ensure full customer satisfaction. In other words, when an organisation develops a physical distribution strategy, it intends to achieve the following objectives:

   i) **Increase efficiency:** Through making profitability by better utilising logistics resources.
   ii) **Improve customer services:** This happens through reducing delivery time, making deliveries more reliable, and lowering the cost to the customer.
   iii) **Increase sales and improve relationships:** Improving the sales strategy through increasing the value of products being provided to customers, and delivering a boost to sales and enhancing customer relationships.

The distribution function within Sales is typically shown as in *Figure 6.1*:

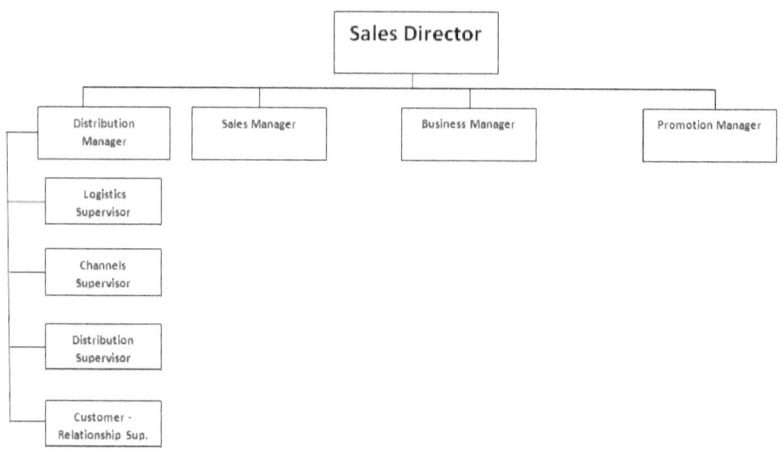

Figure 6-1 The Distribution function as part of Sales

b) **Within Logistics**

Some companies assign all activities of distribution to the logistics department and they take full responsibilities to handle all aspects of the flow of outgoing products to customers. The primary purpose of the group is to arrange and compile necessary goods and/or materials before dispatching them to customer(s) on a timely basis. However, the sales department must have some controls over commercial distribution to ensure striving customer requirements, whereas physical distribution is fully controlled by the logistics or supply chain department. Therefore, the organisational structure chart indicates the distribution function with two reporting lines; one solid line (i.e. direct report to the logistics manager) and another one with a dotted line (i.e. Indirect report to the sales manager) as typically illustrated in *Figure 6.2*.

# Technology Trends in Materials Management          123

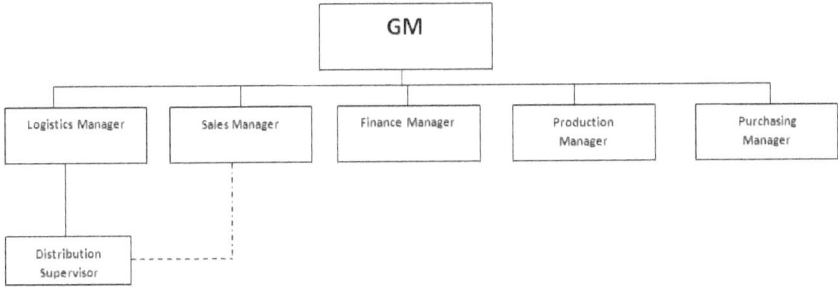

Figure 6-2 The Distribution function as part of Logistics

Success Story

Zeenat Dawani is the Vice Chairman and Managing Director of Deeko Bahrain, one of the manufacturing company in Bahrain. The company is considered a leading supplier of branded consumer disposable and packaging products to food processors, hotels, caterers, and restaurants within Bahrain and other countries in the Middle East. Zeenat began the business in Deeko with one warehouse, few machines, and trucks. The business was initially created to ship small products in and around surrounding small groups of business in Bahrain, however, with her business mentality and long-term thinking, along with a lot of hard work the business in Deeko started to grow successfully from humble beginnings to one of the leading company in consumer disposable and packaging products in Bahrain and the region.

Zeenat realised from the beginning the key role of Logistics as the core of everything in business, and it is ultimately the key to success of the business in order to grow and expand locally and regionally. To this end, she worked very hard to have a very seamless logistics system as a key element in keeping pace with customer demands and outperforming competitors. She thus started to design the logistics strategy in a manner that offers consistency, enhances visibility and streamlines processes so that the new business can meet the new challenges, especially after increasing competition among the producers within the same industry. With a professionally organised business system that setup by her, the company now boasts more than 200 employees working out of different functions, operating as one team in Logistics, Warehousing, Procurement, Finance, Production, HR, and Quality Control working together to meet all demands throughout Bahrain and the Middle East, and they have the ability to answer all business requirements on a real-time base, using advanced ERP and logistics systems.

# PART THREE

# Technology Applications in Material Management

# CHAPTER 7

# Technologies in Warehousing

## 1) Introduction

In the modern world, nothing has been left untouched by technology, and the materials management has become one of the core business continues get developed in the face of technological advancements. Further, Materials Management is always considered as a key to many businesses since it can affect everything from gaining more control over business' inventory, reduce operating costs, and outpace the competition. Therefore, it was necessary for organisations think rationally and strategically to get advantage of technological developments in the area of materials management to face the new competitive market through the new technologies. Warehousing, which is considered the key part in Materials Management, plays a vital role in the overall supply chain and materials management processes to adopt new technologies to create greater visibility over controlling costs, making processes run effectively, and allocating resources more efficiently. Further, a new warehouse involves the use of advanced technology to perform as many warehousing tasks as possible, and many organisations, especially in the west and China, rather than assigning error-prone humans to run the warehousing tasks, they are deploying new technologies to take over. In this chapter, we determine the new technologies that offer to warehousing and then evaluate their effects on the performance of warehousing and their value added to both processes and operations.

## 2) The Importance of New Technologies in Warehousing

There are many objectives for using new technologies in the warehousing tasks and processes, as explained below:

> a) **High Speed Processing:** New technologies can increase speed of processes in warehousing, and of course maximise productivity of warehousing. New technologies not only speed up processes, but improve efficiency, reduce errors, and agility to meet new market changes. For example, the robotic system in warehousing has been successfully implemented in many prominent companies and resulted in increasing productivity based on speeding up processes in machine tending, kitting, and packaging, and it is able to perform repetitive tasks in shortest time. Technologies are also used to speed up inventory management processes through maintaining accurate record items, processing items count, and integrating updated records with other functions. Also, new technologies are implemented in warehousing to speed up processes of carousels, conveyors, stacker cranes, pallet wrappers and other mechanisation. Generally speaking, new technologies can speed up of warehouse operations and has resulted in improving warehouse capacity, quality of work, and the upshots.
>
> b) **Decrease Costs:** Decreasing costs are another huge benefit of using new technologies in warehousing. As stated in previous chapters, warehouse costs represent a large percentage of overall logistics costs along with procurement, and controlling such costs is one of the main objectives of organisations, especially in the manufacturing industry. Although the upfront cost of adopting and implementing new technologies in warehousing is expensive, but the results in realising warehouse costs saving in the long run are obvious. The cost depends on the size and

scope of the warehouse and warehousing, and can range from several thousand dollars to several million dollars. By using new technologies, an organisation can reduce costs of the facility and the process, the equipment, and the software solution that will drive the process, and it includes both fixed (e.g. Equipment & Building) and variable costs (e.g. Labour costs). The following are some key parts in warehousing that can be reduced by using new technologies:

i) **Reduce space:** Smaller spaces are the most cost-effective and, by using new technologies, an organisation can take advantage of storing more in less. According to some recent studies in this aspect, many organisations turning to automated warehouse systems is to save space in a warehouse by making a limited amount of warehouse space which is the only way to optimise movements of inventories.

ii) **Improve inventory control:** New technologies can control inventories more efficiently. For example, automating the picking and movement of warehouse items by computerising all processes will ensure an absolute control over the inventory, and hence an organisation will be able to reduce costs of movements across the whole cycle.

iii) **Protect inventory:** Billions of dollars are lost every year in warehouses. Some of the losses occur because of misplacement or theft, and the rest results from stock damage. By using new technologies, an organisation will be able to eliminate unintentional damage and make a warehouse more secure. For example, using the RFID along with state-of-the-art security systems will ensure accuracy and corrected records, which is considered a means to avoid theft, and making it easier to find misplaced or lost inventory. Also, the bar coding and real-time computer systems employed by a warehouse management system can help ensure that everything is in its place, and track the movement of goods to prevent anything from being lost.

iv) **Optimise cross docking:** Transferring products through loading and unloading processes by using new systems such automated conveyers will reduce huge costs. Cross docking saves both time and money on product management and storage, delivery and shipping, product transference, and labor costs.

v) **Optimise energy systems:** Using automated lighting systems and new forklift machines can reduce energy-associated costs. Also, it will save repair bills and maintenance costs for the automated system against pallet and forklift trucks.

vi) **Reduce labour-related costs:** Labour is an enormous chunk of any warehouse's operating budget, and new technologies are able to contribute in such variable costs through automating the warehousing. For example, using a warehouse management system (WMS), can reduce huge costs by making warehousing run automatically and streamline warehouse processes. Further, streamlining warehousing through automation means that less time is spent in calculating stock and inventory control, and will enable an organisation, thus to measure, plan, track, monitor and report on company's labor assets.

c) **Improve Productivity:** In terms of productivity through automating storage and retrieval machines. Also, they enhance the efficient flow of materials through a warehouse, and that can be integrated the whole systems with other functions like Production, Sales, Financa and Purchasing. One of a good example is WMS, the system enables the warehouse management to find out inefficiencies and address them quickly. Also, the WMS can easily track down how much time an average warehouse operator takes to complete a specific task, and then to measure the performance of the operator and process and hence, eliminate any inefficiencies through automated material handling and management systems. The following are some examples:

i) **Inbound Processing:** New technologies such as WMS can handle automatic receiving with full integration into the purchase order receiving portion of an ERP system. A WMS can also handle directed put away, whereby the system will point warehouse personnel to the bin(s) into which items should be placed upon receipt, based on its pre-programmed knowledge of the warehouse layout. The new systems enhance inventory accuracy immediately after receiving new items in a warehouse, and a barcode system deals with the receipt directly through the handheld as soon as the product hits a loading dock by scanning them into the WMS.

ii) **Outbound Processing:** New technologies are also used in outbound to optimise processes in warehousing. For example, by using handling automatic order allocation, a warehouse is able to locate items for staging and packing, taking care of compliance labeling, automating picking and packing with handhelds, and scheduling dock appointments for shipments. Also, just as critical as inbound processing, outbound processing encompasses picking, packing and shipping orders. Therefore, using new technologies can increase accuracy and speed by eliminating manual steps, resulting in improved productivity through effective product location, and improved customer satisfaction by being able to deliver the right product the first time. A typical WMS increases efficiencies by 10-15%, according to many studies.

iii) **Stock counts:** In stock counts processes, there are many new techniques have been offered for easy count stock items in a warehouse. For example, the "Blockchain" technology (Full details about it is explained under the procurement function in chapter 8) is a system used to keep track of all online transactions securely and anonymously, and keeps all actions made recorded by all computers in concerned functions that are connected into the Blockchain. Furthermore, the system can be utilised in the physical counting process through tracking inventory and documenting the counted inventories and hence creating an indestructible history of the inventories for agility updating in the system. The technology enables radically to reduce time delays and human interventions error that plagues transactions through the traditional inventory systems. Further, all new techniques are linked with a WMS to maintain a complete physical map of the warehouse and is fully capable of directing warehouse personnel to the exact bin locations for stock counts, and hence, productivity is improved by ensuring effective product location.

d) **Stay Competitive:** The competitive advantages concept is associated with the information advantage in the current tough market, and the definition is based on the technology and skills needed to outpace the rivals through knowledge and information being adopted by an organisation. In current and future business environment, there is no choice for organisations to stay in the market without considering new technologies in warehousing. Improved productivity and flexibility while reducing costs are the only way to run the business in the current tough competitive market. To this end, innovative and operational effectiveness strategy is the only ways for organisations to gain a competitive edge and do better than others in the market, and seeking for a competitive edge can't be adopted without considering new technologies. Therefore, automating a warehouse is an important factor to replace some of the traditionally labor-intensive roles and monotonous tasks, freeing up time for warehouse staff to complete other fulfilling and more value-adding roles. With the growing technologies, it's entirely possible that the future, transferring to fully automated warehouse will be required more in order to control every aspect of warehouse operation, including security, receiving, putaway, storage, picking, and dispatch, as well as lighting, temperature control, indirect materials purchasing and even some, if not all, maintenance.

e) **Gain Customers' Trust:** In today's market, technology is the crucial way towards building better customer loyalty towards a business. In order to build customer loyalty through technology, it is crucial for businesses to be interactive with their customers, and hence become as efficient as possible in meeting their customer' requirements based on the modern-day technology. Today, customers have too many choices and their expectations have gone up, therefore, fulfilling their needs is the main concern of management in order to survive, and that can't be achieved without using new technology. To this end, warehouse as a crucial part within the supply chain must be fully equipped with the required technology to meet such objectives, and only technology can fulfill today's consumer needs by having accurate stock records and making a warehouse less prone to errors. Also, new technology is important to integrate with their customers' operations, in order to improve productivity and to provide visibility across the warehousing. Such of this visibility includes information about:

i) Inventory level
ii) Quality of stock items in a warehouse
iii) Stock status
iv) Improved inventory control
v) Warehouse bins status
vi) Warehouse capacity

With such great visibility into the warehouse management, an organisation can improve customer service through such automated solutions and ensure customer satisfaction as ultimate goals.

## 3) Warehouse Management System (WMS)

One of the key technology that is considered to transfer a traditional warehouse to smart warehouse is Warehouse Management System (WMS). Without WMS, other technologies are difficult to work efficiently. To this end, we started to explain the WMS functions as the key pillar to manage other technologies in warehousing:

i) **Basic WMSs:** This type is used to support stock and location control only. The products can be identified by using scanning systems. The system determines the location for storing the received products and registers such information. The storing and picking instructions are created by the system and potentially displayed on RF-terminals. The information for warehouse management is simple and it mainly focuses on throughput.

ii) **Advanced WMSs:** This type includes the basic functions in addition to the capability of planning resources and activities to synchronise the flow of products in a warehouse. These systems focus on throughput, stock and capacity analysis.

iii) **Complex WMSs:** Warehouses can be optimised by using the complex WMSs. The systems are able to provide the information about the location of each product (e.g. tracking and tracing), and they also provide the information about where these products are going to and why (e.g. planning, execution and control). The complex WMSs are able to interface with different technical systems such as, automated storage and retrieval systems (AS/RS), sorter systems, radio frequency (RF), robots and data collection systems. Also, the system offers additional functionality like transportation planning, value added logistics planning, and occasionally simulation to optimize the parameter setting of the system and to improve the efficiency of the warehouse operations as a whole.

## 4) Applications of Technology in Warehousing

The section is separated out into three categories in order to specifically explain each element of technology applied to warehousing. The categories are segregated as follows:

a) Warehouse Robotics Technology, b) Automatic Identification Technologies, c) Communication and Information Technology.

a) **Warehouse Robotics Technology**

New robotics technology has become one of the most sought after technologies for warehouse management. As defined by The Robot Institute of America (1979), A robot is a re-programmable, multi-functional, and manipulator designed to move material, parts, tools or specialised devices through variable programmed motions for the performance of a variety of tasks. The system is one of growing technology being used to carry out a complex series of actions automatically, and used in various tasks and operations, including in warehousing. Today, the robotic technology is considered as an indispensable part of a large manufacturing industries based on its high precision, speed and endurance. Moreover, the technology has become increasingly smarter, more flexible and more autonomous, with the capability to make decisions and work independently of humans.

The term "Robot" comes from a Czech word, 'Robota' which means "forced labor"; and it was first used to denote a fictional humanoid in a 1920 play R.U.R. (Rossumovi Univerzální Roboti - Rossum's Universal Robots) by the Czech writer, Karel Čapek but it was Karel's brother Josef Čapek who was the word's true inventor. However, electronics evolved into the driving force of development with the advent of the first electronic autonomous robots created by William Grey Walter in Bristol, England in 1948, as well as Computer Numerical Control (CNC) machine tools in the late 1940s by John T. Parsons and Frank L. Stulen. The first commercial digital and programmable robot was built by George Devol in 1954 and was named the Unimate, and it was used for transferring objects from one point to another within a distance of about a dozen feet, he then founded a company called Unimation in 1956 to manufacture the robot. After that, the robotic technology was growing and sold to some prominenets companies such as General Motors in 1961, where it was used to lift pieces of hot metal from die casting machines at the Inland Fisher Guide Plant in the West Trenton section of Ewing Township, New Jersey. After that, especially, in 1980s, the robot industry entered a phase of rapid growth. Many institutions introduce programs and courses in robotics. Robotics courses are spread across mechanical engineering,electrical engineering, and computer science departments.

The robotic technology was then developed further as arms of the manufacturing industries to support the Materials Management, especially, in warehousing and distribution to increase the efficiency in preparing, processing, and packaging orders for shipment. There are several logical reasons why the robotic technology has been considered in the logistics and materials management, and one of the utmost reason is to reduce costs and speed up operations in warehousing. In today's market, one of the biggest challenges facing the logistics industry is labor availability, especially, high-quality employees to handle operations in warehousing. Based on the author's experience in logistics throughout his career history, companies can find labors, but not highly skilled labors who can run tasks in a way that could meet the organisational objectives. The following are challenges faced by organisations push them to consider the robotic technology in logistics, particularly in warehousing:

i) **Time management:** This is an important consideration that comes down to having a system ready to fulfill orders and ship them out to customers. Order picking, checking, and processing should not take so much time that it costs money and hence, delay to meet customers' requirements.

ii) **Inventory accuracy:** Stock accuracy can only be ensured via a very sophisticated system. As aforementioned, it is difficult to depend on employees to keep stock items corrected with huge transactions going on in a warehouse. Therefore, information technology is the only solution to ensure all stock items are accurate to meet demands from both production and customers.

iii) **Customer expectations:** Customers have learned to expect more than ever before, especially with availability of required products everywhere through any means of communications. Therefore, in order to retain the customers or to attract new customers, the organisation must ensure shorter shipping times, perfect order accuracy, and amazing customer service. For that reason, new technologies such the robotics system is considered to achieve the customers' expectations.

iv) **Quality of performance:** Organisational success thrives when the right systems are in place, and depending on the traditional methods and thinking related to work performance may not be helpful for an organisation to achieve set goals, especially with the current competitive markets. Today, the only way to retain a business is to consider automated processes to optimise the warehouse management, and using robots are one of the key choices to support the business to improve the quality performance.

v) **Competitive edge:** There is no doubt that a healthy chunk of the company's success will always depend on how innovative the business is, and that of course is the only way to keep a competitive edge on the rest of the market. Using new technology such as robots in warehousing can do things better, faster, and easier, and hence achieving a greater competitive advantage over competitors.

vi) **Costs:** With increasing order volumes, numerous products to navigate, highly personalised order packing and faster shipping requirements, organisations do not have a great variety of ways to control costs of warehousing without investing in the warehouse robotics technology. The robotic system is a solution that enables organisations to effectively perform more tasks in warehousing with less labor and at a lower cost.

**Types of Robotic Systems in Warehousing**

Robots are developed to behave and work like humans, have "Eyes, Hands, Feet, and Brains". In other words, all robots used to support the warehousing will need eyes to see an object, hands to pick it up, feet so that it can move the object to another place, and brains capable of coordinating all these tasks. In this section we will discuss the types of robots offered to warehousing, and how they able to make changes in the warehouse operations.

i) **Open Shuttle:** The Open Shuttles are intelligent, free-navigating precision transport systems, and it is used trackless navigation and directly integrates with complementary systems like Open-source robotics)OSR(. The system is developed by KNAPP company which is famous in warehouse logistics and automation. This type of robot is used to pick a pallet with a certain weight, and transporting it within a warehouse for efficient internal transport, storage and automatic supply to work stations. One of the key characteristics of this type is its ability to react quickly to new routes or changes in transport routes. This driverless system is used to manage all the movements in a warehouse through optimising transport paths, and can prevent any kind of collision.

ii) **Locus Robotics System:** Locus Robots have been developed and deployed a multi-robot fulfillment system that increases productivity and reduces time by utilising innovative work processes. The Locus server guides the autonomous robot fleets throughout the warehouse. The system is mainly used in helping workers pick small items faster and more accurately, thus, it can process more orders faster, using less labor, and with near-perfect accuracy compared to traditional cart-based and follow-bot picking systems. The system can be integrated with virtually any warehouse management system, deploying quickly, and scaling easily as business needs grow.

iii) **Fetch Robotics Freight:** Fetch Robots provide solutions related to material handling through fetch picking arms. The system is acting as the mobile manipulator and helps the warehouse to customise workflows by adding preferred routes, speed maps, keep-out zones and drop-off location so that freights will follow the traffic rules correctly. The system is used in supporting activities in a warehouse through picking items from shelves by auto-arms to their next destination. Fetch robotic systems can improve throughput, efficiency, and productivity with less costs, while working alongside people.

iv) **Scallog Robotic System:** This goods-to-man automated system leverages an autonomous robot fleet, which transports mobile shelving to required picking stations. With Scrollog systems, the operator no longer looses time and effort walking down the warehouse. The system can improve work quality, increase security and reduces physical effort. The operator is more focused on its tasks, thus exposed to fewer errors. Also, it can optimise storage space: an average of 30% space is gained with the solution, using a warehouse space at its best capacity. The Scallog area is organised to allow shelves to fit in a small space. Reduction of movements and simplicity of picking through the scallog system makes flows more profitable.

v) **Swisslog CarryPick Robot:** This system almost does the same function as Scallog systems which turns the traditional picking-process upside-down. Swisslog CarryPick'S is an automated guided vehicle (AGV), used to retrieve mobile racks and then delivers them to manned workstations. A warehouse management system controls the AGVs, which execute tasks simultaneously. With Swisslog carrypick system, the handler no longer needs to move towards the goods, because now the racks become mobile and can drive to the picking-station. One of the main advantages of the system, it allows for an optimal use of warehouse space and saves the pickers tremendous amounts of footwork.

vi) **The Warehouse Drone System:** The system is operated by the operator to perform automatic inventory checks throughout the facility, accurately identifying the inventory in putaway locations, at the frequency of business choosing. Moving the process of information capture into the air provides on demand checks of logistics inventories and avoids the time, expense, and risk of using a people lift to access difficult to reach locations within the warehouse. The system using extensive optical sensors to navigate and identify inventory, and of course determines inventory location safely in a warehouse environment. The power in the drone inventory management solution lies within the sophisticated software capabilities that provides three dimensional mapping, navigation, inventory identification and location accuracy.

## Implementation of Robotic Systems in a Warehouse

Having discussed and understood the types of robots and their main functions in a warehouse, in this section we discuss in which functional areas of the warehouse the robotic systems are installed, and what activities are expected to perform to achieve the ultimate goals in warehousing. The benefits of using warehouse robotic systems in warehousing are numerous, and robots do not drink, sleep, complain, or require a paycheck. With robotic systems, the job can become as easy as possible and the systems can ensure inventory accuracy, provides real-time inventory visibility, and intelligently routes the inventory for efficient retrieval. The robotic system provides full solutions through supporting a range of manual to fully automated receiving, putaway, outbound, and stock counts. The following are solutions that can be gained from robotic systems in the key functions of the warehouse:

a) **Receiving (Inbound):**

   i) Through an automated Guided Vehicle (AGV), fully mobile robots that can transport items from inbound docks to the receiving area, and then to storage, or production, and then take then transport finished goods to storage.

   ii) The system moves along pre-programmed routes to transport inventory from the receiving area to the proper space in the warehouse, then retrieve individual items to fulfill orders.

   iii) The robot takes instructions from a warehouse management system (WMS) via a pick-to-light system, where to pick items from and where to place them.

   iv) The robotic system can quickly load and unload pallets, pick different parts, and perform other functions much more quickly than humans.

   v) The systems are used in the inspection process. Robotic inspection systems perform flaw detection on parts, ensuring the complete part assembly, and measuring parts.

   vi) The systems can print out a label with reference number, date and place in the warehouse.

   vii) The system is programmed to pick materials to their dedicated bin locations without involving human arms.

   viii) Space savings can be achieved with robots since items can be stored in dynamic storage.

   ix) Reduced operational costs can be achieved as the system is linked to the savings in human labour.

   x) The system can improve effectiveness in the receiving function through reducing picking errors and quicker processing times, which result in improved service levels.

   xi) The system allows for an increased flexibility in handling late changes, which could result in a competitive advantage.

b) **Dispatching (Outbound):**

   i) Self-driving vehicles and robots to pull items from shelves and transport them to the dispatching area.
   ii) Robots can schedule movement of the completion of each stage of transportation.
   iii) Robots can Integrate information among other functions.
   iv) Robots look at surface finishes or finding precise dimensions before final loading or after loading items into tracks.
   v) Assemble packages of large items and pallets for delivery to outbound dock after picking them from storage, staging, or picking area.
   vi) The systems are used to reduce walking time and shorten picking routes, integrate with the WMS to ensure accuracy of picking and packing.
   vii) Super-fast in fetching items across the warehouse in all day long compared to human.
   viii) The systems support a safer work environment, since involvement of human is less.
   ix) The systems are reliable based on their impact on throughput and operations for loading correct into trucks.
   x) Easy to pack items before being loaded.
   xi) Save costs in terms of labours and resources.

c) **Putaway**

   i) The robotic systems, through the pick-and-place system, shift and put the pallet to the permanent SKU's location, using a pick to light system fixed in an Automated Guided Vehicle (AGV).
   ii) The robotic systems are developed using an algorithm to evaluate a bin of objects and then figure out how to best grip and organise the item.

iii) Automating material handling operations via robots during the putaway process offers many great benefits such as lower labor costs, higher accuracy, and increased efficiency.

d) **Stock Counts**

i) The robotic systems are utilised to count stocks inside the warehouse using computerised barcode stickers.
ii) The systems used to optimise and balance inventory, aim to solve this problem by automating inventory and routine cycle counts to save time and enhance accuracy.
iii) The systems via RFID and barcode sensors account for the height of an item from the floor, stacked products and depth of the shelf.
iv) The robotic systems on a form of drone with extensive optical sensors to navigate, identify the inventory, determine inventory location, and fly safely in a warehouse environment.
v) The robotic systems can identify the location of stock items physically within a warehouse in near-real time, and can provide regular updated maps of entire inventory automatically on a daily basis.
vi) Using the robotic systems will increase frequency of inventory checks, certainty of inventory levels, reduce losses due to inventory shrinkage, better visibility to out-of-stock, and save on personnel costs, eliminate of resources and equipment, and redirect employees to handle high value activities.

**Benefits of using Warehouse Robotic Systems**

i) **Increase order-to-delivery time:** Robots can facilitate the transport of orders to warehouses, picking robots, the pallet, packaging and pricing departments, loading docks, and shipping containers. The rise of autonomous vehicles and drone delivery concepts will further reduce the delivery times of orders.

ii) **Reduce errors:** Robots can log large data sums and review it for errors accurately. The pinpoint accuracy will reduce reverse logistics processes since all orders will be logged correctly.

iii) **Improve quality:** Warehouse robotic systems perform applications repeatedly and with precision every time. As a result, products are manufactured with similar specifications and processes leading to reliability.

iv) **Reduce cost of warehouse:** Increased use of robots in warehouses means reduced need for human labor, and it is a win for companies looking to reduce their overhead costs.

v) **Better utilisation of floor space:** Automating production line reduces work area footprint. This means that freeing floor space for other processes.

vi) **Reduce workforce burden:** As robots take over most of the physical work in warehouses, human workers will move on to more fulfilling and insightful positions. Robots will also allow people who cannot physically work in traditional logistics operations to gain employment in maintenance fields.

vii) **Reduce transportation delays:** The use of autonomous trucks, driverless vehicles, and drones will reduce shipment delays due to the fast analysis of delivery-impacting factors like traffic conditions, poor tire pressure, and weather.

viii) **Increase safety:** Robots increase workplace safety. This is because workers will move from performing dangerous tasks to supervisory roles where barriers are used to keep operators out of harm's way.

b) **Automatic Identification Technology (Auto ID)**

The second important technology offered to make a warehouse automated is Automatic Identification Technology. Automatic Identification (Auto ID) refers to the identification process of the appropriate stock keeping units (SKU) in a warehouse, using technologies such as RFID, Barcodes, and Voice Recognition. The technology is described the direct entry of data or information in the computer system, programmable logic controllers or any microprocessor-controlled device without operating a keyboard. According to some studies, using Automatic Identification technologies can improve the accuracy of the SKU identification process up to 99.99%, and could reduce errors while performing certain warehousing operations, thus reducing rework and improving time efficiency. In addition to making SKUs accurate, the technology is also used to enforce certain rules such as FIFO, LIFO, and FEFO or a combination of rules that are absolutely necessary for smooth running of operations of inbound, put away, outbound, picking, packing, and stock counts. The most importantly, as mentioned previously, all technologies of Auto ID are needed to be linked to Warehouse Management Systems (WMS) to achieve the fully automated warehousing.

**The Importance of Auto ID in Warehousing**

Auto ID technologies are considered as effective solutions for warehousing. The technology is used to automate the whole processes and to manage data smartly to save time and costs, and reduce errors. The applications of Auto ID are possible for different industries, but they are essential in warehousing and manufacturing as material flow throughout the facility very fast in warehouse and production. The following are the importance of Auto ID applications in warehousing for a sustain competitive advantage:

i) The applications allow companies to improve overall efficiency of inventory management to achieve consumer expectations like speed of delivery, order accuracy, transparency and inventory accuracy.
ii) The applications can locate resources efficiently by supporting warehouse operations effectively in a warehouse.
iii) The applications can accurately meet customer demands by providing the right products at the right time. Avoid any stock out situations in a warehouse.

iv) The applications can raise productivity and boost throughput in a warehouse.
v) The applications can create opportunities for higher revenues.
vi) The applications can achieve an acceptable/positive rate of return on investment (ROI).
vii) The applications offer greater flexibility and effectiveness with less need for human intervention, which is then a means to reduce labour costs in a warehouse.
viii) The applications can help to keep the products valid throughout their storage time.
ix) In case of product recall, the application can discover any defected product quickly.
x) The applications can help to integrate all functional systems in real time. For example, Procurement can place orders based on records being updated via the Auto ID systems in ERP.

**Types of Auto ID Applications in warehousing**

There are different Auto ID applications used in warehousing for stock identification and data capturing, all applications are linked to WMS.

## 1) RFID Technology

RFID is an acronym for "radio frequency identification", it is a form of wireless communication that incorporates the use of electromagnetic or electrostatic coupling in the radio frequency portion of the electromagnetic spectrum to uniquely identify an object in a warehouse, or other purposes. The technology was first offered for commercial purposes in the 1980s, in the transportation industries of railroad and trucking. The technology is developed based on digital data encoded in RFID tags or smart labels which are captured by a reader via radio waves, and the data are then read outside the line-of-sight. The technology works on a real-time information technique to facilitate the collection and sharing of data in a warehouse in order to accurately track the location and status of warehouse resources and to support warehouse operations effectively. The RFID system is composed of three major components:

i) Tag/ Smart Label.
ii) Reader/Interrogator.
iii) Host computer (linked with WMS).

**How it works?**

The technology works through the above mentioned elements as follows:

RFID tags or label contains an integrated circuit and an antenna, which are used to transmit data to the RFID reader (also called an interrogator). The reader then converts the radio waves to a more usable form of data. Information collected from the tags is then transferred through a communications interface to a host computer system (WMS), where the data can be stored in a database and analysed at a later time.

The RFID application can be used in two methods; either automatically or by humans. However, most common applications are access control applications for human beings and linking physical objects equipped with identifiers to databases linked to WMS. Further, the applications for human beings are used to read tags in a range from 3 to 300 feet, for short range transmission. For example, the RFID devices in a warehouse are used to read all SKUs in a bag on the shelf through scanning the bag, but the distance must be within the mentioned range. To this end, two types of RFID applications have been proposed:

i) **Active RFID system:** These are systems where the tag has its own power source like any external power supply unit or a battery. The only constraint being the life time of the power devices. These systems can be used for larger distances and to track high value goods like vehicles.

ii) **Passive RFID system:** These are systems where the tag gets power through the transfer of power from a reader antenna to the tag antenna. They are used for short range transmission.

## Advantages of the RFID Technology in Warehousing

i) **Lower cost & higher productivity:** RFID applications can automate the collection of information about the movement and location of inventories; doing this more quickly, whilst reducing costs and with greater accuracy and reliability than is possible with manual methods and with more detail than can be obtained from techniques such as bar-coding. Data collection can be a by-stock of other activities, eliminating the need for effort in form filling. Identifying products using RFID are quicker than barcode scanning or manual entry of product details.

ii) **Improve quality of Data capture:** Using an RFID approach means data of stock items can be captured rapidly and accurately. Electronic data collection with RFID avoids data transcription errors and avoids "missed items" when used to collect data on large numbers of items at once.

iii) **Reduce capital costs:** RFID help to lower costs by providing better control of stocks in a warehouse. It can help keep track of SKUs in bags, boxes, cartons, etc.

iv) **Better security:** Access control systems using RFID contribute to improve security of business premises. RFID tagging of stock items makes it easier to track inventory "shrinkage" and tags can be used to fight against product counterfeiting.

v) **Increase revenues:** By reducing stock-outs, by avoiding the credibility gap between notional stock available for orders and actual stock present in the warehouse, and by offering improved information on product movements to customers, organisations using RFID can provide a service that creates competitive differentiation and promotes increased customer satisfaction with the opportunities for higher sales and better margins.

vi) **Speed up processes:** The RFID technology is integrated with other manufacturing or supply chain technologies (automated pallet handling, stock picking systems, etc) hence, the speeding up the process cycle from order to dispatch and then delivery.

vii) **Improved regulatory compliance:** Using RFID to control when devices have been inspected or to restrict their movement can form part of a strategy to address health and safety issues or to satisfy insurers or regulatory bodies that processes are being followed.

viii) **Accurate, relevant, current management Information:** Because RFID allows data to be captured in real-time, thus information will be available to other functions such as procurement, productions, finance, etc.

## Disadvantages of the RFID Technology

i) **Too expensive:** RFID is too expensive for many applications as compared to other tracking and identification methods, such as the simple barcode.

ii) **Quality of Tags:** It is difficult for an RFID reader to read the information in case the tags are installed in liquid or metal products. The problem here is that, liquid and metal surfaces tend to reflect radio waves, which makes the tags unreadable. In such applications, they have to be placed in various alignments and angles for taking proper readings, which may be too cumbersome and time-consuming.

iii) **Interference:** This has been observed to take place in RFID systems, when devices such as forklifts and walkies-talkies are in the vicinity. The presence of mobile phone towers too has been found to interfere with these radio waves.

iv) **Standardisation:** RFID signal frequencies across the world are non standardised. For instance, the US and Europe have a different range of frequencies at which RFID tags function.

v) **Consumer Concerns:** RFID is considered by many to be an invasive technology. Consumers tend to worry about their privacy when they purchase products with these tags, as there is a belief that once radio chips are installed in a product, it continues to track a person, and his personal information can be collected by it and transmitted to the reader. So while many stores claim that they deactivate the tags after the product has been purchased, buyers still continue to remain apprehensive of this technology.

## 2) Barcodes

The barcode is another Auto ID technology that stores real time data in warehousing. The technology is simply a means of managing and tracking stock items. The technology is also used via the codes in other purposes such commercial, retails, pharmaceutical, consumer goods, electronics, automobiles, health, etc. The Barcoding technology was first developed in the 1940s, patented in the 1950s, and first implemented by the railroad industry to count rail cars in the 1960s. The technology is a sequence of parallel lines of different thickness with spaces in between. These bars are nothing but the items of information in the codified form, which can be read with the help of a scanner. The information printed in bar code includes, country code, manufacturer name, product details, date of manufacture, material content, etc. These details are required at user end for inventory management.

### Types of Barcoding Technology

Both 1D and 2D barcodes function under the same principle, and both types have applications for which they are individually suited. The following are more details about the two systems:

i) **1D Barcodes (One-dimensional):** Traditional 1D barcodes are linear, a single line of bars encoded in the horizontal width. This type was the standard for many years, and it is still in widespread use today, despite its smaller capacity for information (they can only contain about 20-25 characters, though stacking the characters helps to increase that number). Increasing or decreasing the width of the label changes the number of characters represented. If increased too widely, the barcode cannot easily be scanned. Thus, they can store only a very limited amount of information. Redundancy in the label can be improved by increasing the height of the label. In the event of tearing or abrasions, only a single readable strip of the barcode needs to remain in order for the reader to accurately identify the item that is tagged.

ii) **2D Barcodes (Two-dimensional):** In 2D barcodes, product data are encoded in both horizontal and vertical dimensions using lines, shapes, spaces, colors, and symbols. The systems are designed to encode data differently in patterns of squares, hexagons, dots, and other shapes rather than vertical lines and spaces. The systems store information both horizontally and vertically, resulting in exponentially larger storage capacity; however, an image scanner is required to read 2D barcodes, while a simple barcode scanner can only handle linear codes. The systems can maintain a manageable shape for easy scanning and product packaging specifications. The result is more data, but there is also less redundancy built into the tag. Abrasions or tearing can result in lost data within the tag and possible mis-identification or complete inability to read.

## Language of Barcoding

### Symbologies

The language format between messages and barcode is called a symbology, and it is called barcode standard as well, as it represents the data and information. Code can either in numeric-only or alphanumeric. There have many different types of symbologies for barcode system and used in many different business fields as explained below. Each type of barcode has their unique pattern of bars and spaces to represent the various characters and numbers.

**The Barcode Attributes**

i) **Quite Zone:** Quite zone is an area which allocated to the left and right of the barcode symbol and it is free of printing. The quiet zone is included as part of the symbol and it is necessary for barcode symbol to be read reliably. Besides that, this area provides time for the barcode scanner to adjust to the measurement of each barcode in the message.

ii) **Start and stop characters:** The start and stop characters are a unique character which located at the most left and most right of the barcode symbol. These characters provide reading instruction for the barcode scanner to let it know when to start or when to stop reading. Both characters also allow to be read bi-directionally.

iii) **Data characters:** Data characters also called as "message characters" which contain the encoded information and appear after the start character. Data characters are the information which sent to the barcode printer to print and barcode scanner will decode the information.

iv) **Checksum:** Checksum also called as "check character or check digit". The purpose of the checksum is to ensure that the barcode symbol is accurately decoded, scanned and read correctly by performing a mathematical check.

v) **X-dimension:** The X-dimension is the dimension of the narrowest bar or space in the barcode and usually stated in millimeters. The Barcode symbol usually specifies a minimum value to ensure compatibility between reading and printing equipment. Besides that, X-dimension determines the barcode density, which refers to the amount of information that can be decoded in the barcode in a particular space.

## The most common formats of symbologies

i) **Universal Product Code (UPC):** Universal Product Code (UPC) is the most common code which is numeric-only code and content 12 digit number in the code. 2 digits are on the lower left and right corner is the number system and a check digit. The first 5 lower middle digits are manufacturer number and last 5 lower middle digits is product number that will be created by manufacturer to assign a unique code for each product. This type can provide an efficient, accurate and comprehensive method for conduct food inventory.

ii) **Interleaved 2 of 5 (ITF):** Interleaved 2 of 5 (ITF) is a code that support numeric-only and it can encode two digits, one is in the bars and one is in space. This code only can encode any even number of the digits and if the number of digits is odd, users need to add a leading zero. This code is mainly used in the warehousing application, and in particular to encode packaging level and the trade number is the most common approach using in warehouses. Moreover, the code is common used in shipping cartons.

iii) **Code 39 (Code 3 of 9):** Code 39 (Code 3 of 9) is one of the common barcode in use nowadays, and it can support both numbers and characters. This code is a special character because an asterisk (*) which is used at the beginning and end of the code will create automatically and will not be included in the input data. This code is the most popular in the field of health, manufacturing, automotive, and military.

iv) **Code 128:** Code 128 can support all alphanumeric and numeric only and it has 106 different bar and space pattern and each of the pattern can have one of three different meanings. Code 128 becomes more and more popular compare to others and it was heavily used in today for all the business fields. This code is commonly used in the retail and airline tickets area and in the shipping and retail industry.

## Advantages of the Barcoding Technology in Warehousing

As other technologies, the Barcoding technology has many advantages in improving day-to-day operations in warehousing, and to boost the bottom line in the long run in all areas of warehouse, such as receiving, putaway, replenishment, picking, packing, shipping/manifesting, returns, cycle counts, value-add functions and labor tracking. The following are the main benefits of the Barcoding technology that must be considered as the primary advantages to lead many warehouses to outgrow their old, and make operations in a warehouse more efficient:

i) **Cost effective:** The Barcodes offer automatic and smooth stocks identification via fast recognition and implementation of data in a warehouse. The cost is actually inexpensive to set up a barcode technology, and it can lower overhead and cuts down on training time and labour, along to improving productivity. Further, the technology can lower the costs of capital for carrying excess inventory since knowing exactly what is in stock which helps to avoid ordering an abundance of materials. Also, it can significantly decrease in clerical and paper costs due to reduced need for manual data-entry functions, since transactions can be handled by less employees without papers for printing, writing, etc. The barcoding technology can be used to measure productivity in a warehouse as well which is a means to control the manpower level through comapraing automatically track what work is being done versus to expected output by employees in a warehouse, and hence it is easy to cut costs if the work can be done by less labours.

ii) **Very Simple:** The technology is very easy to use, since it can recall any stock with one quick and simple scan. The technology can be used any employees without a great training program. However, a short training is important to let employees be familiar with the device used for scanning stock in order to avoid any kind of errors when it is being used officially in a warehouse.

iii) **Inventory Control:** With the technology, tracking of stock can become easy and accurately. The data being scanned from portable scanners can be uploaded quickly and accurately to a WMS, and make stock recodes updated in real-time. Thus, the barcoding technology can help to provide accurate stock records to the warehouse and other relevant functions such as procurement, finance, production, sales, etc. Moreover, the technology can empower a warehouse with accurate records during the stock counts processes and to ensure correct stock in the shelves, which allows management to make fully informed decisions that can affect the direction of a company such as, determine correct products for sales, control inventory reorder points, and hence improve customer service.

iv) **Reduce Human Errors:** The Barcods significantly reduce the possibility of human error to the point of nearly eliminating it. The technology as already explained works very fast, more reliable, and less time-consuming than manual data entry thus, it provides 2-in-1 best options for warehousing by quickly and accurately translate data to a WMS and without/less errors, comparing to the manual process. Also, it reduces recognition errors, transcription errors, and transaction errors.

v) **Enhance Security aspects:** The technology is traceable and fully auditable systems, which can provide security tracking, theft deterrence, peace of mind, and a demonstrable reduction in loss/liability via secure a way of encoding information. However, the type of the barcoding system plays an important role to determine the way of security required by an organisation. For example, Code 39 and Interleaved 2 of 5 are known for their high security levels, and companies must have good experience in all aspects of security existed in barcodes before proceeding a purchase order.

## Disadvantages of the Barcoding Technology

Like anything in this life, disadvantages of things are inevitable, and the same can be applied to the barcoding technology. To this end, even though it is efficient and effective in warehousing as explained in the previous section, the technology is not foolproof. The following are some disadvantages of the technology:

i) **Label damage and durability:** The Barcodes that are printed on a torn section of the packaging, or that have been smeared due to humidity or dust, smudged or otherwise damaged, and hence it will present additional scanning problems. If the corresponding numeric code is also illegible due to damage, the checkout process can be significantly delayed while another package of the same merchandise is located and brought to the checkout counter for scanning.

ii) **Discrepancies in pricing:** When changes in prices apply to bar-coded stock, employees in a warehouse or finance may forget to code in the discount price. This, in turn, can lead to confusion and delays in dispatching, and hence inconveniencing the customer who is waiting for the products. In such case, there must be a clear procedure with regard correcting any mistakes in codes as soon as possible, without waiting for long processes which could lead to loss customers.

iii) **Financial and equipment costs:** The barcode scanning device is a pre-requisite for linear barcode scanning process, and thus businesses must be equipped for barcode checkout which could cost the company for implementing the new system that can be prohibitive. Other delays can occur in training employees to adapt to new equipment, and expensive printers must be purchased to print coded labels for any stock that doesn't come prepackaged with a barcode already on it. Dot matrix and ink jet printers, for example, are generally incapable of printing with finely-detailed barcodes.

## 3) Voice Recognition Technology

Voice recognition is another Auto ID technology used in warehousing. The technology is used to perform order selection and other warehouse tasks by using verbal commands that are given to or received from a human. The technology has been used in industrial applications since the 1980's, but became popular in warehouse management after Wal-Mart installed the Vocollect Talkman into their Clarksville, Arkansas distribution center in 1996. The technology can help companies to better serve increasingly demanding consumer expectations while also aiming to improve operations in a warehouse and business profitability as an ultimate goal. The Voice technology is like other Auto ID systems communicated with the WMS using a wireless, wearable computer with a headset and microphone to receive instructions by voice, and verbally confirm their actions back to the system, and the wearable computer, or voice terminal, communicates with the Warehouse Management System via a radio frequency (RF) local area network (LAN). The technology simply works through the following scenario:a voice system tells a worker to go to a certain location, e.g. the employee in the warehouse then confirms the location by saying "XYZ", instead of scanning the location barcode. The voice system then says Pick item "X", and the selctor then confirms the pick by saying "put X", instead of scanning the product barcode.

**Types of Voice Recognition Technology**

The following are types of Voice recognition technology:

 i) **Isolated Words:** The system requires each utterance to have quiet (lack of an audio signal) on both sides of the sample window. It doesn't mean that it accepts single words, but does require a single utterance at a time. Often, these systems have "Listen/Not-Listen" states, where they require the speaker to wait between utterances (usually doing processing during the pauses). The system is good for situations where the user is required to give only one word responses or commands, but is very unnatural for multiple word inputs. It is comparatively simple and easy to implement because word boundaries are obvious and the words tend to be clearly pronounced which is the major advantage of this type.

ii) **Connected Words:** This system type is similar to Isolated words, but allow separate utterances to be 'run-together' with a minimal pause between them.

iii) **Continuous Speech:** This system allows users to speak almost naturally, while the computer determines the content. Basically, it's computer dictation. It includes a great deal of "coarticulation", where adjacent words run together without pauses or any other apparent division between words. Continuous speech recognition systems are more difficult to create because they must utilise special methods to determine utterance boundaries. As vocabulary grows larger, confusability between different word sequences grows.

iv) **Spontaneous Speech:** There appears to be a variety of definitions for what spontaneous speech actually is. At a basic level, it can be thought of as speech that is natural sounding and not rehearsed. The system is able to handle a variety of natural speech features such as words being run together and even slight stutters. Spontaneous (unrehearsed) speech may include mispronunciations, false-starts, and non-words.

**Methods of Voice Model in VR Technology**

All speakers of VR systems have special voices, due to their unique physical body and personality. The voice recognition system is broadly classified into two main categories based on speaker models, namely speaker dependent and speaker independent.

i) **Voice dependent models:** Voice dependent systems are designed for a specific speaker. They are generally more accurate for the particular speaker, but much less accurate for other speakers. The systems are usually easier to develop, cheaper and more accurate, but not as flexible as speaker adaptive or speaker independent systems.

ii) **Voice independent models:** Voice independent systems are designed for a variety of speakers. It recognises the speech patterns of a large group of people. This system is more difficult to develop, most expensive and offers less accuracy than speaker dependent systems. However, they are more flexible.

**Usability of VR in Warehosing**

The VR technology is suited in different operations in warehousing, and the application is mostly used in Order Picking, Stock Checking, Goods Received, Putaway, and package sortation. The VR technology is order selection, and it can handle at least three varieties: full case order picking, split case picking, or the picking of uncartoned merchandise such as garments or tires. The application is also used for replenishment forklift operations. According to some studies, the VR technology can improve order picking productivity by 10% to 20% because the hands free and eyes free operation of the terminal speeds up picking, and trips back to the assignment desk are eliminated. Administrative productivity is also improved by eliminating the work of printing and dispatching picking documents, as well as the task of keying in picking amendments, picking confirmations and catch weights. Further, studies showed that the VR is able to achieve order picking accuracy by 99.9% (one error per thousand picks). The application is also suited for the stock counting processes, whereby the process could improve service levels and reduce time in investigating stock counts and stock discrepancies.

**Advantages of the VR technology in Warehousing**

According to some studies, the VR technology could improve operations in a warehouse dramatically, opening up new possibilities for adding value to the function in terms of productivity, quick picking, and reduces errors. The technology as reported could reduce picking errors between 70% and 90% and short delivery, and operators have gained about 5% to 15% in productivity with voice recognition technology. The following are the main advantages of the VR technology that can be gained by a warehouse:

i) **Increase Productivity:** The VR technology can improve productivity in warehouse in different ways. For example, using the system will keep the picker' hands free as they complete their tasks, allowing them to focus on location and products, rather than on keyboards, displays, or pick lists. This not only increases their productivity, but minimises visual fatigue as well. Also, the technology provides accuracy in processing any transaction, thus maintain control of inventory and manage the fulfillment process. The VR is also considered an effective ceiling operational success by reducing the steps required to complete any task in a warehouse, and improve safety and ergonomics.

ii) **Accuracy:** The VR can achieve more than 99.9 percent fulfillment accuracy of stock on shelves. The technology keeps the right products flowing to keep store shelves stocked and avoid lost revenue. Also, the application allows real-time transmission of inventory information, and just as soon as the stock is moved, inventory records can be adjusted accurately which is valuable in high-volume operations where the timing of replenishment is closely synchronised with order picking.

iii) **Scalability:** The VR solution provides scalability through simple and cost-effective worker and workflow additions, ensuring operations can handle demand while avoiding wasteful investments. The VR is able to support large vocabularies of thousands of words enables it to automate a greater variety of complex mobile work tasks. Further, the application can be scaled to accommodate multiple languages which can be used by workers of their preference e.g. Arabic in case of Bahrain, which enables operations to accommodate diverse labour pools and quickly scale productivity levels.

iv) **Management visibility:** The application allows easy access to the right information by management for faster decisions. Also, the application shows the performance of employees on the floor in real-time, which allows management to gauge their efforts. The VR provides real-time visibility via dashboards into individual worker, group, zone, site and company-wide productivity levels, daily progress and time remaining to complete assigned tasks, enabling management to make rapid operational decisions. Workers are also able to ask for their current performance level, especially valuable in locations using incentive plans.

v) **Flexibility:** The VR technology is very flexible in which it can be used for many different types of applications without any change in hardware. An individual worker might perform receiving with a voice unit in the morning, and then move to order picking with the same equipment in the afternoon. The equipment can be used anywhere in the warehouse where there is a suitable radio signal. It can be used with any product, including items that have no bar codes or are poorly marked. Unlike scanners, voice technology functions well in facilities that have less than ideal lighting. Moreover, the application enables easy integration with other intelligrated system components or existing facility infrastructure. For example, it can be integrated with WMS and/or ERP system, and it comes as a fully-integrated solution that not only provides clear task instructions, but communicates and interacts with multiple enterprise software systems.

vi) **Cost Effective:** The VR can eliminate the costs of warehousing in terms of resources and workers. The application can reduce printing and warehousing documents related to the picking process, and it can reduce movements in a warehouse which is considered costly from operations' point of view. Moreover, the application can help to reduce workers as one picker can handle multiple tasks in a warehouse.

## Disadvantages of the VR Technology in Warehousing

There following are several disadvantages of the VR technology in warehousing:

i) The worker might not hear an approaching forklift when the headset is being worn, and that could be subject to some safety issues.
ii) Pickers could not like talking to a computer, especially workers who were not using any type of systems before launching the VR application in warehousing. Such case is considered as resistant to change.
iii) The system is too expensive, and the cost could be seen as a primary objection to the technology.
iv) There is too much background noise in the warehouse, and that might cause some uncomfortable situations when the VR application is used.
v) The performance of the VR could be degraded by low bitrate codecs, which becomes severe in the presence of data transmission errors or background noise.

c) **Communication & Information Technology in Warehousing**

In this section the focus on a group of ICT applications which could drive value to the whole operations in warehousing through the interface with other internal functions inside the same organisation and with partners in other organisations, using WMS and ERP. We already discussed the function of WMS in the previous section which is the main platform to connect all technologies for improving and streamlining warehouse management process. The main applications being considered primarily for interactions with internal functions and other business entities will be explained in this part.

## 1) Exchange Data Interchange EDI

Electronic Data Interchange is defined as an electronic communication method that provides a standard for exchanging data via any electronic means. The technology is the movement of business data electronically between or within firms (including their partners) in a structured, computer processable, data format that permits data to be transferred without re-keying from a computer supported business application in one location to a computer supported application in another location. This definition includes the direct transmission of data between firms, transmission using an intermediary such as a value added communication network or bank, and the exchange of computer tapes, disks or other storage devices between locations. The latter method fits the definition as data in one business application is computer readable without re-keying by another application even though the medium on which it is stored must be physically transported to its destination. The technology enables to document all business documentation such as transactions related to stock issues, GRNs, Quality verification, stock counts records, and others. The technology is also used between two different organisations, even in two different countries via electronically exchange document (such as purchase orders, invoices, shipping notices and many others). EDI has existed for more than 30 years with different standards such as ANSI X12, Cargo-Imp, EDIFACT, ODETTE etc, some of which address the need of specific industries or regions.

**The impact of EDI in Warehousing**

The EDI technology has a tremendous impact on operations in warehousing such as:

i) Flow accurate information on stockholding.
ii) Speed and reliability.
iii) Strength the business relationship between partners and make it more solid.
iv) Reduces costs in terms of the extensive paperwork.

v) Improves quality information available to other stakeholders within the same organisations, and it could minimise risks of errors.
vi) Creates a comprehensive stock database for valuable market information.
vii) It enables an opportunity for companies to look at their organisations and restructure to meet the new situation.
viii) Makes fully integrated into the overall company systems, particularly those of materials re-ordering, purchasing and inventory control.
ix) The technology can give an extra margin and provides the competitive edge to the successful company.
x) Shorter lead times by cutting the response times to minimum between two partners, and hence decreases stock needs for both partners (buyer & seller).

**Methods of EDI**

i) **Data transfer by magnetic tape:** This involves the normal creation of the transaction records on an online system and the production of a magnetic tape in machine readable form with all the relevant data, which is then physically transferred to the business partner for direct input in his computer system. However, this method is rarely used by some companies due to security issues.

ii) **Data transfer through telecommunications**: This is real electronic data interchange. This involves the processing of the transaction records as in the previous case, but the data is transferred automatically to the business partner's computer where it is entered without the need for any human intervention, which is designed to receive it and where it updates records and other files. The data is transferred via a communications link which can be a leased data line, a dial up line from the public network, a link to a Value Added Network (VAN) or a Trade Community System to which both parties belong.

iii) **Mobile EDI:** This new method processes the transactions similar to the previous one, but with using mobile EDI applications. In warehouses, for example, stock transactions can be updated from mobile devices instead of the desktop PC. EDI data can be accessed through mobile applications and networks and gives a great level of mobile flexibility to document management related to goods handling.

**Types of EDI Technology and Standards**

i) **EDIFACT Standard:** It stands for EDI For Administration, Commerce and Transport. This type provides a set of syntax rules to structure, an interactive exchange protocol and provides a set of standard messages which allow multi-country and multi-industry exchange of electronic business documents. EDIFACT is widely used across Europe, mainly due to the fact that many companies adopted it very early. EDIFACT is considered a framework of regulation for outlining the information in a file data.

ii) **Cargo-IMP:** It stands for Cargo Interchange Message Procedures. This EDI type has been long the number one standard concerns itself with the handling of activities of air cargo.

iii) **ANSI XI2:** It stands for stands for American National Standards Institute. This type is series of EDI transaction sets (documents), divides the transaction sets into market verticals for use in that specific industry or vertical. The ANSI X12 is a universal set of standard documents in electronic business transactions for different industries such as he Warehousing and 3PL industry.

iv) **ODETTE:** It stands for the Organisation for Data Exchange by Tele Transmission in Europe, This type was developed by German automotive companies and spread throughout other manufacturers in Europe. The type is the equivalent of the Automotive Industry Action Group (AIAG) in North America.

The organisation develops tools and recommendations that improve the flow of goods, services data, and business information across the whole automotive value chain. ODETTE has also been responsible for developing communications standards such as OFTP and OFTP 2.0.

**Disadvantages of EDI**

i) EDI uses multiple standards which can often limit how many devices can be connected to the network. The XML web-text language, for example, does not have strict standardisation and that allows for multiple programmers to contribute to the coding.
ii) In addition to rigorous standards, EDI could also have too many rigorous standards bodies with too many document formats which can malfunction in the face of cross-compatibility issues, which you will definitely encounter as you continue to apply more standards.
iii) EDI has a higher price point, which can be a little pricey for new business owners.
iv) Large companies might actually find that EDI can limit the types of partnerships businesses can develop with.

**The EDI Cloud**

This is a new type of EDI implemented via a cloud technology to improve business by eliminating a bulk of the hardware once needed, while also making data access and team collaboration more convenient. The EDI Cloud is a portal to the world of Electronic Data Interchange, and provides an easy and effective means of document transfer among business partners. Through this application, user access can be defined by an administrator so that each user can be granted access to only the parts of the cloud that are relevant to them. For example, sales might just see the order summaries, where as, warehouse staff might only create the pallet or carton labels, and others might control the product lists.

**Advantages of the EDI Cloud**

i) **Work from anywhere**: Cloud technology allows business to work from anywhere with an internet connection. This means an employee can collaborate on projects while traveling, access important documents outside of the office, and even work from home.
ii) **Disaster recovery**: Businesses rely on data more than ever before. Although it may seem unlikely, a catastrophic event or even a simple system failure can destroy businesses physical hardware, resulting in a loss of data stored on that hardware. However, data stored in the cloud is regularly backed-up and disaster-proof.
iii) **Flexible costs**: With traditional IT systems, businesses have to invest large chunks of money up front to get a system that will allow last a while, meaning business end up paying more than you need. Most cloud services allow businesses to pay for what businesses use now and increase the capacity when they need it.

## 2) Enterprise Resource Planning -ERP

Enterprise Resource Planning (ERP) is the advanced software solution to enable integrating of transactions oriented data and business processes throughout a firm. The system was evolved by the Grtner Group in the 1990s, and it was evolved from MRP, a term that is well-known in business, which already explained earlier. The logistics function is one of the key modules in ERP that manage and process data related to suppliers, products, orders and deliveries, and inventories. The ERP technology is a software used to link business functions in an organisation, and leads to increase the efficiency. The technology can offer high value to an organisation whose aim is smooth planning and controlling of stock management to achieve long term profitability and maintain a solid competitive edge. The technology is useful for management through integrating and automating operations in warehousing by giving accurate information for better decision making.

## Basic functional modules of ERP in Warehousing

i) **Sales and Distribution:** This module is part of the large logistics area. It supports customer relationships through the purchase order to invoice. The module is closely integrated with inventory management and production planning.

ii) **Material Management:** The module is responsible for the coordination of planning, sourcing, purchasing, storing and controlling materials.

iii) **Production Planning:** This module is used to manage the life cycle of the product manufacturing process as per customer requirements.

iv) **Quality Management:** This module is used to manage planning and controlling product quality through quality data of raw materials, manufacturing products, WIP, and finished goods.

v) **Plant Maintenance:** This module is used for managing inspections, determining technical condition of equipment, and preventative measurements.

## Benefits of ERP in Warehousing

i) **Efficient Storage Administration**: New warehouse management functionality permits important indicators to be monitored within the ERP framework. As a result, complex warehouse operations can be easily managed and optimised. The latest modules include radio frequency technology, integrated shipping, work-in-progress tracking, batch and serial number management, returns management, and many other features, indicating a new standard for warehouse operational transparency and accuracy within a facility. Focusing on real-time stock control, special functionality has been added to allow manufacturers to identify exactly what is in a warehouse and balance the inventories of multiple storage facilities at any given time.

ii) **Increases productivity:** A complex warehouse management ERP system with advanced warehouse management functionality can help businesses get more out of their activities. Operatives will complete more tasks during the same time frame, which means better availability, quicker turnarounds, and higher return on investment.

iii) **Reduces costs:** An ERP system not only simplifies warehouse operations (e.g. workers will be able to verify inventory levels without doing any physical checks, managers can coordinate activities with just a few clicks, etc.), it also ensures greater accuracy. There is always a risk of data redundancy when multiple separate systems are used. This can lead to errors that will cost a lot for business. With ERP, information can be updated in real time across the entire company, so there are no discrepancies that may cause errors. Eliminating errors and rework will lower business operational costs.

iv) **Customisation:** The technology enables to customise processes for better fit operations. For instance, businesses can choose to integrate the warehouse management module with the quality control module to prevent storing the items that have not passed the quality test in the warehouse. Specific features can be enabled to use cross-docking functionality, setup company policies for sales and transfer orders to and from warehouses, allocate batch/lot numbers for better warehouse management, etc. More modules and features can be added later on, as needed.

v) **Better Customer Service:** Having a sophisticated warehouse management ERP solution in place can ensure businesses always have sufficient quantities of products in stock. As shown above, it can also help reduce redundancies, errors, and rework, which will result in better customer service.

vi) **Enhances Collaboration:** ERP helps an organisation to have a control over all the suppliers and distributors. This creates the ability to know what they are doing all the time, since ERP is able to bridge the gap among stakeholders. With ERP, all stakeholders across the network can share vital information like demand, forecasting reports, inventory levels, status of production, transportation plans and many more in real time.

**Type of ERP Technology**

There are two primary types of ERP solutions available to businesses as follows:

i) **On-Premises ERP**: Is deployed locally on hardware and servers, and managed by IT staff. Businesses that choose this option want greater autonomy over their implementation.

ii) **Cloud ERP**: Software looks and works the same as traditional ERP. The only difference is how it is deployed. With cloud ERP, instead of hosting servers and storage hardware on-site, an ERP provider hosts this for business. Cloud-based ERP brings enterprise-grade security to protect today's businesses, as well as a lower cost of ownership, ease of use, and configuration flexibility. It also provides business with real-time access and visibility into business information via a cell phone or other mobile device from everywhere in the world.

**Disadvantages of ERP**

The following are some disadvantages of ERP need to be considered by businesses:

i) **Costly:** ERP can cost more than less integrated or less comprehensive solutions.
ii) **Resistance:** It can be rejected by some employees, which could lead to divert management attention.
iii) **Unnecessary dependencies:** Integration of truly independent businesses can create unnecessary dependencies.
iv) **Training:** Training requirements could take resources from daily operations.

v) **Mammoth task:** Harmonization of ERP systems can be a mammoth task (especially for big companies) and requires a lot of time, planning, and money.

## 3) Internet of Things - IoT

The IoT is a new interconnection of technology heralded as the next industrial revolution, implying radical change, disruption, and an entirely new paradigm for the planet. The technology has a great impact on the economy by transforming many enterprises into digital businesses and facilitating new business models, improving efficiency and increasing employee and customer engagement. This technology, which was resembling back to the year 1988, starting with the field of ubiquitous computing. In 1991, Mark Weiser framed his ideas for the computer of the 21$^{st}$ century, is simply a network of dedicated physical objects (things) that contain embedded technology to communicate and sense or interact with their internal states or the external environment. The connecting of assets, processes and personnel enable the capture of data and events from which a company can learn behavior and usage, react with preventive action, or augment or transform business processes. The IoT is a foundational capability for the creation of a digital business through three things measure and report data, which are as so called "A"s. Aware i.e. sense something, Autonomous i.e. transfers data automatically to other devices or to Internet services, and Actionable i.e. integrate some kind of analysis or control.

### The Impact of IoT on Warehousing

IoT integration has a large positive impact on warehousing, as it allows stock levels to be monitored in real-time, making inventory management in high volume businesses more efficient. The following are the key advantages of IoT in warehousing:

i) **Strength Communication:** IoT enables the communication between devices in a warehouse. For example, an IoT is able to connect all of the devices communicate across a common platform in a warehouse. Within the platform resides all of the information that is generated by the connected warehouse, providing a holistic view of the warehouse. In a simple scenario, an IoT can indicate when a forklift whose battery needs to be charged or conveyor whose ball bearings need lubrication or replacement. The technology is able to analyse the data in a near real-time manner, improving the performance and efficiency. Changes to the performance of the system can be made to achieve a specific business goal.

ii) **Automation and Control:** All devices are communicated automatically based on sufficient control without human intervention, and the process can be run faster and on timely manner. For example, sensors in a warehouse can detect locations of people, vehicles and merchandise, and can transmit locations and events to warehouse managers and management systems. Also, employees in a warehouse can use smartphones to load boxes or pallets on a truck in the correct order for dispatching.

iii) **Reliable Information:** Through an IoT, reliable information can be obtained for making better decision by management. For example, management uses an IoT to collect the data and analyses it to find patterns and correlations. In simple scenarios, and technology may report the low level performance of employees and accordingly, management can take the decision to make a corrective action to adjust it quickly.

iv) **Better Monitoring:** An IoT can alert workers or managers of low stock, identifying the temperatures and other conditions at which inventory is best kept fresh and valuable, monitoring the conditions at which workers are safest and most productive, determining the best layout and configuration of the warehouse, finding ways to maximise profits with inventory, scheduling worker shifts, improving the safety of the workers and improving the security of the inventory and warehouse among many others.

v) **Saving Time:** Using an IoT will save the amount of time in performing operations in a warehouse. For example, an IoT can save time and reduces operational mistakes when arranging cars for delivering products. In another scenario, the system helps in tracking real-time accurate information about inventory and prevents out of stock situations immediately.

vi) **Reduces costs:** An IoT can save money through stripping down the warehousing costs, and that is done in different ways. For example, reducing manpower, machines, and other ineffective resources that might be decided to cut.

**Applications of IoT in Warehousing**

The following are the key applications of IoT in warehousing:

i) **Apply to track materials around a warehouse:** The application is used to track pallets from the time they are packed till the time they are moved to Production or final delivery to consumers. All activities throughout the movement of pallets could be known ahead of time.

ii) **Apply to truck Drivers and Deliveries:** The application is able to truck both drivers and deliveries to ensure goods are delivered to the required location precisely. It can indicate if any odd case occurs, such as lost or misplaced shipments.

iii) **Apply to monitor the condition of goods in a warehouse:** The application can monitor all items in a warehouse to avoid damage or loss. For example, some warehouses are used to store perishable or sensitive goods that only stay fresh for a short time, and/or require specific environmental conditions to maintain quality. With an IoT application, the employees can monitor the humidity in the storage area, the temperature of the product, and the shock and vibration levels these items experience during shipment.

iv) **Apply to improve the safety of equipment:** An IoT application is used to improve safety of equipment in a warehouse. By placing sensors and cameras on various equipment, the system can track and measure the use of equipment. For example, the application is able to measure and analyse data recorded to predict any maintenance that might be needed for any machine or equipment e.g. Forklift. Also, it can be used to monitor the usage of equipment, conducting regular preventative maintenance checks, and installing cameras are all ways that warehouses have been able to improve safety conditions.

v) **Apply to improve inventory control:** The application is used to provide managers with instant insight into how effectively stock items are being controlled in the warehouse. Inventory sensors and tags allow for real-time visibility and inventory control. Real time visibility monitors the exact location of any item and prevents inventory from getting misplaced. Sensors also help to automatically and accurately update the stock levels of warehouse inventory.

vi) **Apply to improve space allocation:** The IoT application is used to optimise space in a warehouse. The technology can ensure the stored goods are retrieved, processed, and delivered as quickly as possible. Therefore, no space issues will occur in the warehouse since all goods are moved smoothly, which means all spaces are well utilised, and hence no problems in the throughput process.

## Disadvantages of IoT

There are some disadvantages of IoT still need to be overcome:

i) **Lack of Standardisation:** There is no common standard of compatibility for the tagging and monitoring equipment and machines in a warehouse or other facilities. Thus, it could be hard for devices from different equipment and machines to communicate with each other.

ii) **Complexity:** Since the application is run in a vast and diverse network, a single failure in either the software or hardware can have disastrous consequences.

iii) **Privacy and Security:** Because all machines and equipment are connected to the internet, which means information is readily available, it thus makes it harder to keep confidential information out of the hands of hackers and other unauthorised users.

iv) **Increase jobless:** Automating systems mean many employees, especially unskilled ones could loss their jobs, and hence it leads to unemployment issues in the society. This is a problem with the advent of any technology and can be overcome with more education.

v) **Technology addiction:** Technology addiction is damaging our personal and work lives, and basic human interaction skills will be reduced across society.

# CHAPTER 8

# Technology Applications in Procurement

## 1) Introduction

Another element of Material Management that takes the responsibility of the process of buying materials, parts, services, etc., for an organization. As explained in chapter 5, they take the role of identifying and selecting a supplier, negotiation price, ordering, expediting, receipt and payment. After the advancement in technology, digitalising processes have become necessary in the procurement function for giving a boom to the processes in order to act as quick decision enabler and to become a key driver of next-level performance and results. Moreover, the procurement is considered one of the key functions that affect business factors in terms of leaking out costs and hence reducing net profits, since they consume a certain amount of resources such as employees, time and money of an organisation, which is counted in general as the non-liable assets and it affects the growth of an organisation in the current tech-arena. To this end, the new technology has given a boom to the procurement process to become more innovative in handling the processes to improve the business strategy through creating appropriate structures and make use of suitable instruments and advanced technology such as E-Procurement, which typically describes business-to-business purchases that are done online or over some digital network or platform.

In this chapter the author explains the new technologies that are able to enhance the procurement processes, communications with stakeholders, and lead the function to achieve organisational goals and objectives.

## 2) e-Procurement

Procurement departments 'role is to constantly engage a variety of stakeholders, both internal and external. So we focused on the use of technology in Procurement from the electronic information and communication perspective with both internal and external stakeholders, which is considered the key to achieving the strategic goals of an organisation.

### Definition of e-Procurement

Electronic procurement is considered the key technology used to support operational, tactical and strategic procurement. e-procurement which was spawned from Enterprise Resource Planning (ERP) systems in the late 1990's, nowadays works based on the ERP concept which already explained its characteristics in warehousing in the previous chapter. An ERP in procurement, however, takes a different form as the automating system to run the whole processes of procurement either internally with other departments or externally with suppliers or distributors using the internet based applications and technology. To this end, e-procurement is formed as a platform to transfer processes to fully automate in order to communicate information simply and efficiently, streamlining the global procurement process, and reducing time and costs without compromising on standards and quality.

### E-Procurement Forms

E-procurement is supported by various forms of electronic communication, and it is used in both intra and inter organisational systems. The following are the main forms of e-Procurement:

i) **Electronic Data Interchange:** Inter-organisational information system using structured data exchange protocols often through value added networks.
ii) **e- MRO:** Mechanism for ordering indirect items from an on-line catalogue.
iii) **Enterprise Resource Planning:** Automation of procurement related workflows, including auto-faxing, auto-emailing or other forms of messaging directly with suppliers.
iv) **e-Sourcing:** Way of identifying new sources of supply using internet technologies.
v) **e-Tendering:** The process of inviting offers from suppliers and receiving their responses electronically.
vi) **e-Reverse auctioning:** Using internet technologies bidders usually bid down the price of their offers against those of other bidders until no further down-ward bids are received.
vii) **e-Auction for disposals:** Using internet technologies for on-line auctions of items for disposal.
viii) **e-Informing:** Use of internet technologies for gathering and distributing procurement related information.
ix) **e-Collaboration:** Collaborative procurement related planning and design using facilitating technologies.

## e-Procurement Advantages

There are countless benefits in using e-Procurement within the procurement function, and the following are the main ones:

i) **Reduces Transaction Time**: individual business transactions can be completed much more quickly; they are not restricted by office hours and may not even need human intervention, thus increasing the capacity to complete transactions on a real-time basis. This means that downstream processes are not constrained by waiting for transactions to be completed.
ii) **Electronic Catalogues**: the development of e-Catalogues has enabled organisations to market their product offer

electronically, this has been a fantastic marketing tool for sellers and for buyers, there is price transparency (buyer can easily see how much items cost) and buyers can hence compare offers from various e-Catalogue vendors.

iii) **Increases Standardisation**: With the electronic catalogues mentioned, there has been a move by some suppliers to offer a more standardised offer, thus allowing buyers to easily compare the offers from e-Catalogues, however care must be exercised in these comparisons as it is difficult to assess the quality of products without samples. In case of any doubt, buyers can request samples and take time to make their own assessment.

iv) **Wider Spread Supplier Bases**: Because the virtual e-Procurement portals are web-based, buyers can search suppliers worldwide, meaning a wider selection of products and services are available to the organisation, meaning that when items are not available locally it is still possible to source these. It is important to remember the time and cost of shipping goods, but it's great to know that it is possible to source items from somewhere in the world!

v) **Simplifies Global Procurement:** With the e-Procurement applications supporting various languages, currencies, international taxation and financing, shipping regulations and more, it is simple for buyers and suppliers in different countries worldwide to communicate and cooperate.

vi) **Increases Productivity**: As e-Procurement automates some of the procurement and wider business processes typically handled by employees, this will free up time for the team to spend on more strategically significant functions and tasks. For example, with automated matching of invoices, goods can be ordered, processed and paid in a matter of minutes; the key however is to ensure that the supplier is set up with the buyers' systems support as much automation as possible.

vii) **Simple Configuration and Scalability**: e-Procurement applications can be configured to suit the individual needs, i.e. the buyer and the supplier, and can grow with the organisation as needs be. It is important to select suppliers for both the current requirements as well as the possible future need, so gaining an understanding of the technical infrastructure development plans of suppliers will help buyers to select possible long term partners.

viii) **Creation of Trading Communities**: Because the e-Procurement application is internet based, it allows for both vertical and horizontal trading communities to be developed. This means buyers can consolidate buying power and it also opens up opportunities for new supply chains. The opportunity to consolidate the requirements of smaller buyers via consortia or trading communicates has enabled smaller business to access prices historically reserved for bigger buyers, thus fuelling a fast developing SME sector. Many Chambers of Commerce and other local business organisations operate such buying communities.

ix) **More Cost Efficient**: With the time reductions and increased supplier selection, development of trading communities, more opportunities for purchasing surplus goods and services at below market price, and much more, it isn't surprising that e-Procurement proves to be much more cost efficient than traditional procurement.

## E-Procurement in Internal Level

It is also important to understand how an e-Procurement is impacting on intra organisational systems as this is just as important as inter organisational systems. Internal processes are a vital part of the logistics and e-Procurement, and plays an important role at the level of customer service within the organisation, which could also have a bearing on the level of external customer service. The following are the significant determinants of e-Procurement internally:

i) **Improves quality of services:** The system enhances quality of works conducted among the internal groups in an organization.

Through an integrated and real time process, all transactions are processed smoothly with less effort and errors.

ii) **Transparency:** The system provides transparency in any transactions take place between Procurement and other departments through an integrated system, and that can eliminate corruption in an organisation.

iii) **Monitors transactions:** The system enables monitoring of all transactions during the procurement process, providing functionalities such as the approval hierarchy which is an end to end process that facilitates the procurement process from planning to payment. Further, e-Procurement provides features that include approval hierarchy through the transaction life cycle.

iv) **Improves customer service:** The system can improve customer services by providing accurate information to Procurement about the stock level in the warehouse, which enables Procurement process purchase orders quickly before any stock is running out.

v) **More collaboration:** e-Procurement can deliver a collaborative procurement through driving the standardisation of best practices in procurement, which can help an organisation become more efficient, improve process management and save money in the long run by sharing procurement knowledge, expertise and experiences.

vi) **Reduces Costs:** An e-Procurement system is able to save huge dollars by improving internal and external processes. Using such the collaborative system can improve overall ordering efficiency and allows a company to better track where its spending is going. The system doesn't rely on personal relationships as much and find savings by searching thousands of vendors at once to find the best price for a company. Also, the system can ensure all expenditures are pre-approved by an authorised person and according to budget, and that will obviously reduce expenditure overall and ensure best discounted prices with suppliers.

## Types of Transactions Conducted Internally in e-Procurement

a) **Forecasting:** The system processes of production and sales forecasting automatically via:

   i) Avoids running out of stock for critical components.
   ii) Calculates the optimal quantity to reorder based on reordering level.
   iii) Assesses the most profitable products and suggest them as a priority purchase.
   iv) Suggests which inventory should be cleared of stock and no longer purchased because of a significant drop in demand.
   v) Determines better plan purchase of seasonal products.

b) **Stock flow:** The system can organise all transactions related to stock flow activities through automating requisitioning, automating workflow, eliminating double handling and less paperwork, and reducing cycle time which results in less inventory. This process is always running between Procurement, Warehouse, and Production.

c) **Inventory Control:** The system can control inventories through the following steps:

   i) It can define the limits of inventory planning, for each item, by specifying an item's minimum and maximum allowable quantities, and item reorder quantity (to send alerts and notifications when reaching these limits, and to generate automatic triggers) The purchasing module that generates a request based on the result of the re-order point checks.
   ii) Manages the inventory-related transactions like disbursements, receipts, retirements, transactions between different inventories, transactions within the same inventory, returned items, and custody transactions.
   iii) Manages the inventory cycle count, and have the ability to automatically generate corrective transactions.

iv) Manages the inventory's information from basic data (like inventory titles, type, custody and warehouse), assigned managers, assigned users, and other advanced information.

v) Manages the inventory planning and shipping schedules, including consolidated forecasts and release information with certain restrictions.

d) **Budget control:** The system enables to control budget through stopping maverick spending and stay on budget. The budgetary control is done between Procurement and Finance as all transactions are run based on real-time and full visibility into spending management, and using the e-Procurement helps the financial controller to see real-time budget impact and hence determine the level of control and predictability through liaison with Procurement.

e) **Vendor Rating:** Procurement can evaluate supplier performance in an efficient way by obtaining information being entered into the e-Procurement by the warehouse, quality, or/and distribution. For example, based on information entered by the mentioned departments respecting the delivery performance, lead time, quality, price, or some combination of variables, the purchasing analyst can easily assess each supplier accurately.

f) **Communicative means:** The procurement department can easily communicate with other departments using the email notification mechanism of the e-Procurement system. The system supports a dedicated notification being dispatched to other users when a specific action is performed, e.g. PO being placed, expediting process, approval notification, etc. Through the notification mechanism, the users are kept informed for all progresses being made.

## E-Procurement in External Level

In this section the main roles of e-Procurement from the inter organisational perspective, and thus it is important to underscore the importance of the use of Internet-based technologies, as they play the key role of interaction between Procurement and suppliers. Basically, e-Procurement refers to the B2B, B2C, B2G, P2P, etc. The new technology like e-Procurement plays an important role in the e-Commerce world (e.g. B2B), and it forms about 80% to 90% of the entire business process according to some studies. Through the B2B, many buyers and sellers are collaborating smoothly, dealing with one-stop shopping, and business deal is conducted in fixed and pre-negotiated prices. However, B2B is conducted through both e-Commerce and e-Business as they do the same electronic services in terms of e-Buy and e-Sale, as both systems include the internet and electronic data interchange (EDI), and they are sometimes used interchangeably, as long as they address the same processes, and share the same technology infrastructure of databases, application servers, security tools, systems management and legacy systems. Also, both systems involve the creation of new value chains between a company and its customers and suppliers, as well as within the company itself. However, there are some differences between the two systems in some aspects as explained below:

a) **e-Business:** The system includes e-Commerce functions as well. e-Business strategy is more complex, more focused on internal processes as well as external processes, and aimed at cost savings and improvements in efficiency, productivity and cost savings. The following are some examples of e-Business:

   i) **Auction sites and classified sites**: e-Business helps people to sell merchandise via the internet, such as companies like eBay.
   ii) **Software and hardware developers**: These companies develop new ways to network businesses together, like Microsoft or Adobe.
   iii) **Digital marketing**: This e-Business is based on selling products or receiving commission through promoting products on the internet. For example, if a YouTube personality talk about a product and offers a discount or referral code, that is one aspect

of a digital marketing business. Many vloggers and bloggers have been able to parlay this aspect into a profitable career.

iv) **Setting up online storefronts**: Owning a website that supports other businesses is an e-Business. Companies like Wix and Squarespace make starting an online company "plug and play," meaning they offer templates to register a domain name, craft a website, choose colors, generate marketing reports, accept payments, and keep track of inventory. All sorts of add-ons can increase how much this type of e-Business makes.

**There are two types of e-Business, which are:**

i) **Pure-Play**: The business which is having an electronic existence only. Example: Hotels.com.
ii) **Brick and Click**: The business model, in which the business exists both in online i.e. electronic and offline i.e. physical mode.

b) **e-Commerce:** This system covers outward-facing processes that touch customers, suppliers and external partners, including sales, marketing, order taking, delivery, customer' service, purchasing of raw materials and supplies for the production and procurement of indirect operating-expense items, such as office supplies. The technology, which was developed in early 1990's, involves new business models and the potential to gain new revenue or lose some existing revenue to new competitors.

**Features of e-Commerce**

i) **Non- Cash Payment:** There is no cash in e-Commerce transactions, buyers can use credit cards, debit cards, smart cards, electronic fund transfer via bank website, e-Wallets, and other modes of electronic payment.
ii) **Online Service Availability:** e-Commerce automates the business of enterprises and the way they provide services to

their customers, and it is available anytime, anywhere, and can be accessed by any person.

iii) **Improved Sales**: By using e-Commerce, Sales can be improved by providing anytime and anywhere without any human intervention, and 24*7 open shopping opportunity to consumers.

iv) **Support:** e-Commerce provides various ways to provide pre-sales and post-sales assistance to provide better service to customers.

v) **Communication improvement:** e-Commerce provides fast and reliable communication with customers and partners.

vi) **Inventory Management:** e-Commerce provides inventory reports faster and accurate when required, as products are easier to maintain. Thus, inventory management has become very easy to control and process.

## Advantages of e-Commerce in Procurement

The E-commerce technology provides advantages for both buyers and suppliers as listed below:

a) **Advantages for buyers:**

   i) **Easy to compare:** In e-Commerce, there is side by side comparisons are readily available and easy to do. There are many alternatives available to compare products in terms of quality and specification in order to obtain the best suit one for buyers.

   ii) **Time saving:** There is no physical effort with e-Commerce. The buyer can make navigation in minutes and that will save a lot of time to obtain required products.

   iii) **Convenience:** Every product is at the tip of buyer's fingers on the internet. The buyer can get a product through a simple search engine on the internet.

   iv) **Easy to find reviews:** The e-Commerce allows reviewing any product by past users and thus, the buyer can be aware about the positives and negatives of product before proceeding a transaction to buy.

v) **Easy to get more verities:** With the help of e-Commerce, the buyer can have more verities and choice of goods.

b) **Advantages for Supplier**

i) **Rise in sales:** With the help of e-Commerce, products are sold all over the world. In other words, products are available everywhere without thinking of market or product segmentation as in physical markets.

ii) **Available in 365 days:** e-Commerce is available in 24*7, 365 days. Thus, suppliers have more opportunities to offer their products without bearing the overhead costs of labour and resources.

iii) **Expand business reach:** In e-Commerce there are great tools to sell products via the internet. For example, products can be advertised in different languages and online communications with buyers. Thus, it is easy for buyers to get full details about any products from the supplier accurately.

**Disadvantages of e-Commerce**

i) **Security Issues:** Even with advanced technology, there is still fears over some security issues within e-Commerce. For example, there is still a chance of hacking cards at the time of payment of bills, and such perception can't be avoided without having full confidence about the e-Store buyers are interested in.

ii) **Bandwidth issues:** The bandwidth still remains an issue through the e-Commerce business. Bandwidth is the amount of traffic space in a website, and e-Commerce sites need to be careful of exceeding their bandwidth as do businesses that get a lot of high-profile media coverage.

iii) **Product issues:** In e-Commerce there might be some major problems in terms of damage to the product, the misplace of product, receive different product, etc. These can be problems for both buyers and suppliers as products in question must be returned back to the supplier which impact on costs and business reputation.

## Models of e-Commerce (e-Procurement)

As explained earlier, referring to the key models of e-Commerce as the main applications of e-Procurement which is moving corporate procurement to the World Wide Web through allowing employees of a buying organisation to purchase goods and services and allow suppliers to manage and communicate the fulfillment of the purchase orders submitted. In this section we discuss the main models of e-Commerce that form the real relationship between stakeholders throughout the entire procurement cycle.

a) **Business to Business - B2B**

Business-to-Business e-Commerce is defined as the buying and selling of goods and services between companies online, and it is used in inter-organisational coordination, communication, and results in achieving cost savings and competitive sourcing opportunities for the buyer organisation. The model is also known as C2M (consumer to manufacturer) in case of dealing with a manufacturer via e-Commerce, and it is used by Procurement to cover two types of purchases; direct and indirect;

i) **Direct purchases:** It involves materials, such as raw materials and components, which go into the finished products sold to the customer, and it is conducted based on company-wide standards and controls.
ii) **Indirect purchases:** It involves goods and services that are not part of the finished product, but support the internal business

activities. For example, items such computers, office equipment, operating supplies and office supplies. It also involves a wide variety of items of different complexities, and caters to a range of internal needs and preferences. The process through indirect purchases are conducted based on a decentralised system, which is considered as incompatible applications within the same organisation.

**The model has two key features**

i) **Flexibility in pricing**: Transactions between businesses via a B2B procurement system provides variability in the pricing of products between purchasers, and that concept of haggling is rare in other models e.g. C2B.
ii) **Integration of business systems**: To realise increased productivity and savings, businesses involved in B2B will integrate their internal systems together, enabling less human intervention.

**Impact of B2B Procurement**

The obvious impact of B2B procurement concerns the impact on firms involved in the procurement activities within an industry. The ability to digitise many of the activities associated with procurement has fundamentally changed this aspect of many industries. The following are most affecting the realisation of the value of B2B procurement:

i) **The search power:** The model helps the buyer to control costs through the power of the search engine in the system, and thus the buyer can locate an appropriate seller who offers optimised prices for required products. The search costs process conducted by the buyer can be categorised in two situations; i) when he/she looks for a supplier for contracting purchases, and, ii) when the buyer looks for the appropriate product to order. In both situations, the B2B model via associated search engines are able to considerably lower the search costs, which can be quite significant in large organisations, and it helps the buyer to easily search using multiple methods to ensure that he/she can find the right product even with limited available information. This "user-friendliness" of the system reduces the "premium buys", where the user goes around the procurement system and incurs higher processing and product costs.

ii) **Optimises processes:** B2B procurement involves electronic document routing and information flow, thus reducing labor costs involved with manual processing. It enables automatically route the product request for the necessary approvals and order placement with suppliers. This reduces the transaction cycle time and gets the materials to the user faster. As the system requires minimum data inputs during the information processing cycle, along with the ability of the buyer to eliminate many of the sources of errors, which thus results in lowers the cycle time, errors and the processing costs.

iii) **Monitoring and Control:** Using the B2B procurement will lead to achieving twin objectives of responding effectively to the user needs as well as leveraging their combined buying power. For example, the buyer can search the catalog to identify the most cost-effective supplier and place their orders, and hence management can aggregate the demand for the whole enterprise and use this to negotiate competitive prices for the products, which they can then make available to any business unit, irrespective of the size or location of the unit. Centralised control, combined with the availability of an increased range of items in the electronic catalog, motivates more users to order through the e-Procurement system, reducing the extent of premium purchases. Thus, the major benefits of the model monitoring and control are reduction in average product price and reduction in premium buys.

iv) **Coordination:** Using the B2B procurement can provide real-time information flow and is less costly to coordinate with suppliers and users. This leads to faster resolution of any problems and results in lower order cycle time. Lower communication costs of the system and the lesser time spent by the procurement staff in coordination results in lower transaction costs. Improved coordination capability also helps to speed up product development cycle time and avoid design of duplicate components.

**Advantages of B2B Procurement**

The following are the main advantages of B2B procurement, which are important to achieve the strategic goals of an organisation:

i) **Many options:** B2B procurement is the only option to meet a multitude of suppliers for all business e-Sourcing, Catalog Management, and procurement needs. Through the system, the supplier and buyer both can reach agreements over any deal with respect to products and prices, customised offers, reliable delivery, and make safe payments. Such approach streamlines the process of buying and makes it more personalised.

ii) **Ease of Automated Processes:** Switching from conventional paper-based procurement to an automated process can save business productive time that the procurement team can spend on operational tasks instead of tedious purchase processes.

iii) **Seamless Inventory Management:** The e-Procurement suite of Purchasing Platform gives the business the option of linking inventory real-time with the business profile, and it will give a complete control over the inventory, which in turn will enable business gain better control over the inventory and considerably reducing inventory and storage costs.

iv) **Invaluable Data Insight:** It empowers Procurement to perform all tasks and purchase decisions based on a paperless system.

v) **Market Research Locus:** The B2B procurement gives advantages to Procurement to conduct a rigorous market research by showcasing a vivid database of vendors and products in its catalog marketplace.

vi) **Aggregation Of Buying Power:** The B2B procurement provides various features such as employee Spending Limits, Purchasing Transparency, and Management Approval on Orders. It can also offer aggregation of buying power completely.

vii) **Builds Strategic Relationship with Suppliers:** The B2B procurement is a means for building a strategic relationship with suppliers. Both parties have opportunities to reach agreements that suit best for their business, and they can have a mutual long-term contract through a B2B marketplace that will bind both parties in any business deal.

viii) **Better Sales: The** B2B e-Procurement is a collaborative approach in the supply chain management process to increase loyalty, and it helps the buyer to be aware about all products being offered by the supplier in an easy way, and it is an approach to motivate the buyer on upselling and cross-selling opportunities.

ix) **Lower Costs:** Due to an effective supply chain management process, the B2B e-Procurement is able to lead to lower costs for the businesses. In most cases, the work is done through automation that eradicates chances of errors and undue expenditure.

x) **Data Centric Process:** One of the main advantages of the model is that it relies on effective and factual data to streamline the whole process. In this way, errors can be avoided and proper forecasts can be made. With an integrated data-driven approach, businesses can calculate detailed sales statistics.

## Disadvantages of B2B e-Procurement

There are some disadvantages of B2B e-Procurement that are considered flaws for businesses:

i) **Limited Market:** Compared to the B2C model, the B2B model has a limited market base as it deals with transactions between businesses. This makes it a bit of a risky venture for small and medium that rely on e-Procurement.

ii) **Lengthy Decision:** The B2B e-Procurement involves lengthy decision in any business deals since there are two businesses involved. The process may involve dependence on multiple stakeholders and decision makers.

iii) **Inverted Structure:** Compared to other e-Commerce models, consumers have more decision making power than sellers in the B2B e-Procurement. They may demand customisations, impose specifications and try to lower price rates.

iv) **Technology Skills Required:** The B2B e-Procurement requires buyers to have skills in using and dealing through the model. Recruiting a buyer must based on their skills in performing PO's via WWW, along with his/her skills in procurement. The new trend could increase costs in terms of higher pay might be asked by a new buyer who is highly skilled in both procurement and technology.

v) **Lack of Privacy/Security:** Some businesses (Suppliers) don't want their pricing to be visible for all the world (including their competitors) to see. They may also be concerned about exposing their customers' credit card numbers of possible security breaches.

vi) **All the Eggs in one Basket:** The B2B e-procurement could let Procurement only rely on the websites. Even just a few minutes of downtime or technology hiccups can cause a delay in performing transactions which could cause substantial loss for business and affect on customer services.

b) **Business To Consumer - B2C**

The second model related to e-Commerce (e-Procurement) is B2C (Business to Consumer), which is considered a business through the internet between businesses and consumers. The model, which was first utilised by Michael Aldrich in 1979 who used television as the primary medium to reach out to consumers, provides relief to wherever people are coming to all relevant information about offered products and services, along with introducing an innovative platform between applied to organisations that indulges in selling its goods or services to the consumers over the internet medium. The model contains information regarding the goods and services offered, which stored in a database and is represented as online catalogues.

**Characteristics and Process of B2C Model**

The B2C model consists of various elements as explained below:

i) **Cataloguing:** The cataloguing is the process of displaying items from a database based on categories and sub categories selected by the clients, and it includes the description, the quality, price, and brand of the item. Through the B2C model, the product catalog is created for the buyer to view and purchase through a simple process. Cataloguing is a major operation used in the business to consumer e-Procurement, and for that reason sellers or suppliers are always required to ensure the technique is fulfilled all needs of the consumer.

ii) **Planning and Generation Orders:** The order planning and generation enables initiation of individual orders as well as block orders in an easy to use and flexible manner. The model enables any order goes through a complete life cycle, and each stage is indicated by the system throughout the cycle.

iii) **Cost Estimation and Pricing:** Before setting the pricing, there must be a complete study of cost estimation. Pricing is the bridge between customer needs and company capabilities. Pricing at the individual order level depends on understanding the value to the customer that is generated by an order and evaluating the cost of filling each one. After ordering, estimation of cost can be done first and then set the pricing procedure.

iv) **Order Receipt and Accounting:** When price setting is over, accountability of the products and their cost is maintained. It is the basic and a major step in the accounting system. After an acceptable price quote, the customer enters the order receipt and entry phase of ordering. Order receipt is necessary for the billing of the different products.
v) **Order Selection and Prioritisation:** After setting the price of the product, the major operation is order selection and to set the priority for the selection of the final goods. Customer service representatives are responsible for choosing which order to accept and which to decline.
vi) **Order Scheduling:** This step is used for a healthy customer environment. In this phase prioritised orders get slotted into an actual production or operational sequence.
vii) **Order Fulfillment and Delivery:** After completion of order scheduling, the next step is to fulfill and deliver the order. During this phase the actual provision of the product or service is made.

## B2C Categories

There following are five categories of the B2C model:

i) **Direct Sellers:** Direct sellers, such as online retailers, sell a product or service directly to the customer via a website. One can further divide direct sellers into e-Tailers and manufacturers. e-Tailers are electronic retailers that either ship products from their own warehouses or trigger deliveries from other companies and stocks. Product manufacturers use the internet as a catalog and sales channel to eliminate intermediaries.
ii) **Online Intermediaries:** Online intermediaries perform the same function as any other broker. The business allows non-B2C companies reap some of the benefits. Brokers offer buyers a service and help sellers by altering the price-setting processes.

iii) **Advertising-based Models:** Popular websites rely on advertising-based models. These websites offer a free service to consumers and use advertising revenue to cover costs. They draw a large number of visitors, making them ideal advertising streams for other companies. Advertisers need to pay a premium to sites that deliver high traffic numbers.

iv) **Community-based Models:** Community based models combine the advertising method that relies on traffic in the sites that focus on specialised groups to create communities. Community sales and advertising take advantage of social and network marketing by focusing on specific groups that want specific products.

v) **Fee-based Models:** Pay-as-you-buy or paid subscription services fall under fee-based models. The most common of these are online subscriptions to journals or movie sites such as NetFlix. These companies rely on the quality of their content to convince consumers to pay a usually nominal fee.

## Advantages of B2C model for Procurement

i) **Safe:** The buyer can login to a secured part of web site to conduct purchases. All activity conducted by Procurement is protected by the Secured Sockets Layer (SSL) protocol, which provides business and the customer with peace of mind.

ii) **Easy to Navigate:** The buyer can review products and services which required by business, and check details about products such as their price, delivery terms, and specification, etc. Also, buyers can easily browse the website for products they need and add them in the cart before initiating the checkout process.

iii) **Short Buying Cycle:** With the B2C, the buyer can buy any product within a shorter buying cycle, and procurement can proceed any purchase via the B2C faster than B2B which is subject to workflow and long order approval, therefore resulting in a much longer buying cycle. The B2C thus, is useful in purchases urgent or small value items that required less procedures.

iv) **Accurate Inventory Visibility:** The B2C model is useful to view the 'in/out of stock' status, and it is typically suffice for B2C end users.

v) **Collaboration:** The B2C facilitates automation of electronic transactions between enterprises, support real-time exchange of information enabling a collaborative process.

vi) **Customisation:** The model provides the ability to offer products and services in real time, the ability to customise goods to the needs is greatest.

vii) **Transparency:** The B2C model is the trend set to making purchases and billing in more efficient and transparent. Companies are utilising digital procurement platforms to centralise purchasing and to get rid of error-prone and time-consuming manual tasks in their procurement processes.

viii) **Global Reach:** The B2C model is the global reach it has, and it can be processed from everywhere to offer products to customers. This availability to sell to anyone and anywhere makes sure success is inevitable to both seller and buyer.

ix) **No Physical Overheads:** B2C has advantages by lower overhead costs for both buyers and sellers, as all transactions are conducted virtually via the website, companies can only make purchases by one buyer through the online store.

x) **More Data available:** The model opens the door to more information about sellers or suppliers through using analytics tools like Google Analytics, and that will help the buyer to discover demographic information about the supplier.

xi) **Trackable Suppliers:** The buyer is able to track the supplier's business success via online. Reports through Google Analytics can show how products are liked by customers before proceeding any orders.

**Disadvantages of B2C**

i) **Security:** This might be the main concern to go through B2C by procurement. Such concerns about the security of B2C pose a serious limitation to the model. As mentioned earlier, trust is a Must in such businesses before making any business deal with sellers, since perceptions regarding security will remain a limiting factor in the B2C model.

ii) **Infrastructure:** Such services only work with the existing infrastructure, but if such facilities get subpar, buyers will get problems accessing to the B2C, and other models of e-Procurement. So it is important to remedy such infrastructure gaps, otherwise it stands as an ongoing limitation.

iii) **Competition:** Of course competition in such new trend is very tough, and the field of B2C e-Commerce consists of fierce competition for the eyes and dollars of consumers. Thus, businesses wanting to sell online must compete with entrenched B2C eCommerce giants, such as Amazon and Staples, which brought billion of dollars in the last 10 years from the B2C businesses. Businesses must also capture market share from other small vendors, many of which offer identical or nearly identical products or services. Vendors that sell custom or specialty products, may face somewhat less competitive conditions due to the unique nature of their products.

iv) **Limited Interaction and Customer Service:** Unlike shopping in a brick and mortar store or talking directly to a service provider, the B2C places inherent limitations of interaction. At the product level, customers must make decisions based on images, product descriptions and reviews. Customers cannot handle a product to see if it feels good in their hands or weight enough to indicate the manufacturer employed quality materials in its construction. Much of the customer service provided by those engaged in the B2C happens strictly through digital means, such as forms on the website or an email, often with long lag times between filing a complaint and receiving a reply. Thus, the system could create a dissatisfied environment for customers, which result to think of the traditional marketing.

c) **Business To Government - B2G**

The third important e-Procurement model that we should discuss as part of e-Commerce is B2G, it is also known as Business-to-Administration (B2A). The importance of this model because the model covers the business between Procurement in the public sector and businesses in the private sector, and for that reason it often referred to as a market definition of "public sector marketing". The process through the B2G is business-to-government is conducted through a concept that businesses and government agencies use central websites to exchange information and do business with each other more efficiently than through a traditional method. Also, the designation is also used in any relationship between the subject of public administration and the enterprises as one of the basic e-Government models. The model offers a tremendous e-Procurement service in which businesses learn about the purchasing needs of government agencies, in turn request proposal responses. The model works based on the procurement cycle based on the initiative comes from a government organisation and businesses towards one objective, and that works in a virtual workplace in which a business and a government agency coordinate the work on a contracted project by sharing a common site to coordinate online meetings, review plans, and manage progress. The model is usually used the ICT solution that converts such communication to the electronic form or to describe a solution that simplifies communication between the public administration and enterprises (e.g. Internet portal of the procuring authority or electronic solutions for purchasing). The most important in this model that B2G interaction is considered to be more complex than B2C or B2B models since it deals and interacts with specific governmental organisations, which might have specific procurement procedures and regulations.

## Characteristics of B2G

The following are the characteristics of the B2G in terms of achieving the mutual goals for both the public sectors and private sectors:

i) Provides a high degree of information security by using digital certificates to identify the users.
ii) The model normally initiates a B2G relationship through a platform by identifying its needs to the public. This is done through its yearly budget, Requests for proposals (RFPs), Public sector organisations (PSOs), Request for information (RFI), Request for quotations (RFQs), Sources sought and suppliers respond to them, and other types of solicitations.
iii) The model works based on a government project that is publicised in order to receive bids or proposals from businesses interested in winning the contract.
iv) The model is also used by third party websites which collect government contracting opportunities, and then initiates deals with private sectors.
v) The processing of B2G contacts requires a considerable amount of time and resource capacities in companies which in return leads to higher costs of bureaucracy. By optimising B2G processes, monetary and efficiency benefits can be achieved for both sides: companies and public administrations.
vi) Web-based purchasing policies in the B2G are more clear and transparent in the procurement process than other models, and it also reduces the risk of irregularities.

## Advantages of B2G

i) The model can help companies in the private sectors to be validated as an industry leader and prize contractor.
ii) The model enables companies to elevate and differentiate their businesses from competitors for new contract awards, as

well as General Services Administration (GSA) schedule and Governmentwide Acquisition Contract (GWAC) task orders.
iii) B2G can increase company's visibility with government buyers, including enhancing their social media presence.
iv) B2G can help companies to open a new avenue for two-way exchange of information with government buyers, potentially enhancing the relationships with government buyers.
v) The model can create more opportunities to influence government acquisition decisions.
vi) The model can help companies win more recompete, extensions, new contract awards, and task orders.
vii) It accentuates the team's value, expertise, insight, and experience.
viii) It emphasises the value of products and services being offered by companies.
ix) The model is able the procurement practitioners assess the private sector's performance, which is important to gain the trust to their products and services in the market.

**Disadvantages of B2G**

i) The B2G model is still considered insignificant among other e-Commerce models, as government e-Procurement systems still not implemented in many countries, especially in developing countries. It is only used in some developed countries like USA, and some countries in Europe.
ii) The process might take more time through the B2G model than other models based on the government licensing procedures which are more complicated compared to procedures in the private sector.
iii) B2G corruption is thought to be one of reasons to keep the model undeveloped in many countries. One of the examples is a phenomenon in the Chinese construction sector, especially in public construction projects, which resulted in many accidents and losses.

iv) Compared with B2B error and corruption, B2G errors and corruption could receive widespread public attention, because rent-seeking government officials who regulate the market can abuse their power to bypassing laws and regulation benefits at the expense of the whole society.

**Some Examples of B2G portals are:**

- B2GMarket
- Bidmain
- ScanPlanet.com
- SupplyCore

d) **Consumer To Consumer - C2C**

The fourth model of e-Commerce related to Procurement in business is C2C. The C2C model, which is also called person-to person (P2P) e-Commerce, is a new breed of e-Commerce through a means used by one consumer to sell goods or services to another consumer online. Such the system has been in existence for a long time before even emerging the Internet, through advertisement in newspapers, but it was developed with the growing technology through the global connectivity provided by the Internet. The purpose of the C2C system is to enable consumers to sell directly to other consumers without having to go through a middleman. However, the system is usually facilitated by a third-party site that helps take care of the details of the transaction. Using the C2C model will allow the seller to keep more of their profit and the buyer to potentially purchase the goods at a better price. Procurement can have a good opportunity to find some goods through C2C as its market is projected to grow in the future because of its cost-effectiveness, the cost of using third parties is declining, and the number of products for sale by consumers is steadily rising. Therefore, buyers can consider it as an essential business model to get the right products, the right quality, and with less price. Further, the C2C model has become one of the popular business in recent years through the social media and other online channels, so buyers can approach any supplier in an easy way.

**Types of C2C Business**

The following are three types of the C2C model with examples:

i) **Providing Full Procurement Cycle:** In this type of C2C business, sellers provide full services to buyers, which included shipment and delivery. For example, Craigslist is an e-Commerce platform that connects people advertising products, services, or situations. Craigslist not only provides a platform for buying, selling, and trading products, but attract buyers through posting monthly classified ads, such as employment opportunities and property listings. Like many others, this platform requires the seller to ship items directly to the buyer.

ii) **Providing Guidance and Advise:** This type concerns with advise & guidance, for example Etsy allows company owners to create their custom website on which to market their products to consumers. The C2C site offers guidance and tools for growing a business that ranges in price according to a company's stage of development. There's also a Sale on the Etsy application that helps to manage orders, listings, and customer queries efficiently.

iii) **Provide Items based on the Auction Method:** In this type, sellers provide the auction option along with the fixed price option. For example, eBay features two types of product listings: fixed price items and auction items. Fixed price items can be purchased quickly by selecting the 'Buy It Now' button. Auction items feature a 'Place Bid' button for entering bids and show a current bid price. These items are open to bids for a predetermined time and are declared "sold" to the highest bidder.

## Advantages of C2C for Procurement

As mentioned, the C2C model is a phenomenon of online sales that has widespread in recent years by attracting companies to focus on it to benefit from its popularity based on its simplicity and efficiency in the new technology era, which could be done through simple channels, including the social media. Using C2M or C2C means the company directly connects to consumers, cutting off the flow of fare-increase links and removing intermediate links, so that consumers can purchase autonomous customisation and low-cost high-quality products. To this end, the procurement as well can benefit from such the model for achieving the strategic goals in different ways. The following are some advantages of C2C based on the e-Procurement:

i) The C2C business platforms can offer buyers with the advantages of being able to influence the economic terms of the transactions and attaining merchandise, which may not be available in the other channels.
ii) The C2C platform is reachable at any time, thus the buyer can effortlessly consider it for items required urgently, thus they save time looking for the item they need online without wasting time going around.
iii) Buyers can benefit from using the C2C model through bidding in the auction process to achieve the minimum price of products, which are available at higher price in other channels.
iv) The C2C model can be used to select the best sellers, most popular products that are available in the buyer's area hence to benefit from low delivery costs.
v) The buyer can benefit from many features of C2C, such as: i) simple searching process, ii) distribution costs, iii) inventory costs. Hence, buyers can keep their costs low and get a higher margin.
vi) Through C2C, the buyer also cuts off the intermediate links, including the dealers, inventory, operations, warehousing, taxes, profits, etc., and only need to consider the organisation's profits.

vii) The buyer can benefit from the payment method through the C2C model. e.g., the use of online payment systems such as PayPal.

**Disadvantages of C2C**

i) One of the biggest challenges of C2C e-Commerce that can be faced by the buyer is the payment method, even though it has some advantages as explained above. For example, If the seller doesn't get paid for the item that they sold and shipped, it can be hard to get the item back or make the buyer payment.
ii) Another disadvantage in the C2C model is about the quality of purchased items. For example, if the buyer doesn't like the quality of the item, the seller isn't always required to offer a refund like a store would.
iii) Lack of trust is also considered as a disadvantage of C2C. It is not easy to convince an organisation, especially big ones to conduct businesses with entities or persons under uncertain environmental states, which they might think kind of risky situations.

e) **Direct To Customer - D2C**

The last model we will discuss "in brief" in this section is D2C. This model is considered a new innovation in the field of the supply chain, founded in 2016 by a company "Unilever" in the USA. The D2C is defined as a low barrier-to-entry e-Commerce strategy that allows organisations, especially manufacturers, to sell directly to the consumer. Thus, manufacturers can bypass the conventional method of negotiating with a retailer or reseller to get products on the market, they customise the experience to accurately reflect their products. In other words, the model goes against a lot of traditional retail expectations through occupying a lot of space in a larger retail store, companies now think purely of online businesses. The D2C is like other e-Commerce models relies on an online medium to reach customers. The main characteristics of this model are listed below:

i) The D2C is a means of a direct connection between the manufacturer and the final customer.
ii) The model is best suited to enter the market directly, without a middle man entity.
iii) The D2C model is able to enhance some logistics processes through making a direct connection of manufacture-side and demand-side without a delivery or logistics stage.
iv) The model is useful and practical in terms of making mass customisation for smart supply chain management (SCM). In other words, the model can help two companies provide customised services at a low cost after understanding customer's taste and demand and making data.
v) The D2C has advantage to connect directly to consumers, and it helps to eliminate the barrier between the producer and the consumer, giving the producer greater control over its brand, reputation, marketing, and sales tactics. Plus, it helps the producer directly engage, therefore, learn from their customers.
vi) The D2C model has a good advantage to collect consumer data upon their consent, so that companies can constantly improve themselves and the customer experience.
vii) The D2C is a way to understand the importance of physical experiences through allowing customers to practically test a product in order to have a full picture about their preferences, and that can improve the ongoing relationships with customers based on the trust and reliability.
viii) A good example of D2C business is practiced by an American multinational corporation "Nike". The company has expanded its brand to tap into some common D2C practices. Instead of relying on customers to buy Nike products online or at Nike standalone stores, the company has made an offensive push to get products to consumers. Nike stores in Los Angeles offer

curb-side pickup, and other stores have vending machines full of Nike socks. Customers are more willing to buy things when they're convenient and fun.

ix) Finally, the D2C strategy is able to hinge on providing a convenient service from start to finish, and companies thus, can have full control over their products and sensitive data about their customers to provide an exceptional level of thoughtful service.

## 3) Mobile Procurement - m-Procurement

In the previous section we detailed about the e-Procurement platform and some key models that is included, and in this section we explain a new technology which provides the same basic functionality as e-Procurement, but through the mobility aspect platform. Mobile procurement (m-Procurement) is the technology used by the procurement function to streamline their procurement process from a mobile device. The technology has emerged after improving cellular networks to 3G and then 4G, and 5G is expected to take over by some prominent companies, such as the Samsung Galaxy S10 5G, the Huawei Mate X, and the LG V50 ThinQ. The m-Procurement technology, which was first depicting in 2003, known as performing the procurement process over the internet using mobile devices, and the term is used to describe the use of mobile application to support all stages of the purchasing process.

**Definition of m-Procurement**

Mobile procurement is the business to business purchase and sale of supplies and services over the internet using mobile devices. The term "m" is used to describe the use of mobile devices, wearables, wireless networks and internet connections to streamline and enable procurement activities. The technology allows to perform the following tasks in Procurement:

i) Performs procurement processes between buyers and suppliers at remote locations.
ii) To make bidding using personal digital assistant (PDA), a smartphone, or other emerging mobile equipment such as dashtop mobile devices.
iii) To improve communication and collaboration between buyers and suppliers.
iv) To perform processes faster and easier.
v) To leverage software-side servers to move data along.
vi) To provide capabilities to complete the procurement process from start to finish, using all features such as searching for items and services, comparing vendors and prices, submitting purchase requests, approving requests, electronic signing, purchase orders and invoicing.

## Technologies Requirements for Developing and Implementing m-Procurement

a) **Mobile Agents**: The software is an emerging and exciting paradigm for mobile computing applications, it has the following features:

   i) It helps to design a wide range of adaptive, flexible applications with non permanent connections by adding mobility to code, machine based intelligence, improved network and database possibilities.
   ii) The program is autonomous and can move through a heterogeneous network under its own control, migrating from a host to host and interacting with other agents.
   iii) It decides when and where to migrate.
   iv) It can execute at any point or suspend its execution, move to another host and continue its execution on that host.
   v) The software is autonomy, mobility, goal driven, temporarily continuous, intelligence, cooperation, learning, reactivity etc.

vi) It simplifies the development, testing and implementation of distributed applications because of their ability to hide the communication channels and show the computation logic.

vii) It can distribute and redistribute themselves throughout the network and can act as either clients or servers depending on their goals.

viii) It can increase the scalability of the applications because of their ability to move work to an appropriate location.

b) **Service Oriented Architecture (SOA):** It is a software architecture pattern in which applications or systems are constructed from underlying (and usually distributed) software services that conform to a specific set of characteristics. The primary goal of SOA is software development agility, i.e. the ability to respond the change easily, and cheaply, thus allowing businesses to rapidly respond to changing markets. The following are the main features of the SOA model:

i) It is an architectural model for designing software systems.

ii) The main-objective of SOA can be formulated as enabling an efficient and powerful service orchestration for the implementation of new applications.

iii) The method is used for distributing, heterogeneous and interacting software applications.

iv) The basic building blocks in the method are autonomous and platform independent software units (services), which are amply specified by separate interfaces.

v) These interfaces allow for the publication, search and orchestration of the described services within open and heterogeneous networks like the Internet.

vi) The main idea of the method is to have all components in a distributed application communicate and interact via services.

vii) The method provides a platform for an efficient and effective publication, discovery, binding, and assembly of these services.

viii) Finally, the software is implemented as web services, i.e. they operate over the ubiquitous web HTTP protocol, and are implemented either using XML-based SOAP or the lightweight (and more popular) REST paradigm.

c) **Cloud Computing:** Cloud computing is a model for enabling ubiquitous, convenient, on-demand network access to a shared pool of configurable computing resources (e.g., networks, servers, storage, applications, and services) that can be rapidly provisioned and released with minimal management effort or service provider interaction. The main objective of Cloud Computing is for the network to independently provide software, services and computing infrastructure, and it can be thought of as an infrastructure that provides storage, processing and applications as a service. These services can be accessed over the internet by using some standard browser.

d) **Mobile Cloud Services:** Mobile cloud computing use cloud computing services to deliver applications to mobile devices. These mobile applications can be deployed remotely using speed, flexibility, and development tools. The cloud applications can be built or revised quickly using cloud services. They can be delivered to many different devices with different operating systems.computing tasks and data storage. Thus, users can access applications that could not otherwise be supported. The best example in this regard are: the Google Application Engine and the Apple Cloud, which are the most prominent cloud providers that enable the consumption of cloud services from the handset, since their cloud solutions are completely integrated for their own mobile platforms. Some open source that enables the communication with multiple clouds include: Jets3t, Jclouds, Typica and Amazon Navite API. The Google Application engine contains all the set of services provided by Google. It uses a SaaS approach for delivering services over the Internet. It also provides support for the storage (Google for developers). The Android mobile platform is tied to the solutions provided by Google, and thus most of the services released over the Internet were extending for proving a mobile version 8.

**Advantages of m-procurement**

The following are the most important benefits mobile technology can provide to improve the procurement visibility and insight:

i) The m-procurement empowers the buyers with a smartphone or a tablet to work from any location, and that give them satisfaction and motivation to perform their jobs such as in finding the best prices and the fastest delivery time, etc. In other words, the technology through its mobility provides a platform that gives instant access to share information.
ii) The m-Procurement can help managers to approve items while they are away from their desk. This in turn reduces the time required to process approvals and requisitions. This is especially relevant in industries that often operate remotely, or in the factory or field, such as manufacturing and mining.
iii) Using the m-procurement devices will generate a financial benefit in the form of cost savings, greater employee productivity, and greater operational efficiency.
iv) The m-procurement can be used in conjunction with a mobile device's GPS. For example, if an employee is at a vendor's store or office and would like to take a picture of an item, a smartphone or tablet with a camera can accomplish that task instantaneously.
v) The m-procurement can manage the mobile inventory availability and the urgency of the service call, and thus improving internal customer satisfaction because of faster emergency response times.

vi) The m-procurement can improve accuracy in terms of material record time, invoices updated, and enhancing cash flow. In other words, it can enhance the process between warehousing, finance, and procurement.

vii) The m-procurement can manage and control fuel consumption through having an updated record mileage, which is an indicator of fuel purchases for vehicles used by other departments. Therefore, the buyer can keep close track of mileage which helps to control fuel spending more efficiently.

**Disadvantages of m-Procurement**

Main disadvantages that come with the use of m-Procurement in business include:

i) Adoption of m-Procurement is considered a challenge and management must consider many factors before allowing such the technology to replace the existing methods. For instance, management must have clear answers for the following questions: How they manage and configure it? What data plan will you use? When will a solution provider offer mobile access? Will it improve the processes? Will the employees accept it?

ii) Security is considered a big concern in any technology. The m-Procurement is prone to cybercriminals, as well as a bigger risk of employees mishandling critical data. For that reason, management must insure mobile procurement initiative doesn't put the business at risk, and ensure the security team is able to lock down and manage the devices and the data being transmitted, especially, the data stored in a cloud storage.

iii) Costs are one of consideration in implementing m-Procurement through adopting technologies related, devices, training, and maintenance, which management needs to allocate a huge amount of funding.

iv) The m-Procurement is possible to distract a workplace, as the range of technologies and devices increases, so does the potential for them to disrupt productivity and workflow in the business.

## 4) Blockchain Technology in Procurement

Blockchain is a relatively new technology first implemented in 2009 by a coder named Satoshi Nakamoto. Basically, the blockchain technology is an unchangeable ledger of events or tasks recorded as transactions and ordered by time. Every transaction is therefore fully auditable. This means that there is always a link between transactions to combat any double-spend attempts (spending the same dollar twice). Let us quote the definition from Seebacher & Schüritz, which gives a clear picture about the technology: "A blockchain is a distributed database, which is shared among and agreed upon a peer-to-peer network. It consists of a linked sequence of blocks (a storage unit of transaction), holding timestamped transactions that are secured by public-key cryptography (i.e., "hash") and verified by the network community. Once an element is appended to the blockchain, it cannot be altered, turning a blockchain into an immutable record of past activity."

As shown in the above definition, the Blockchain technology consists of two key parts; Block and Hash. When we say the words "block" in this context, we are actually talking about digital information (the "block") stored in a public database (the "chain"), and each Block stores information that distinguishes them from other blocks, e.g. like an ID number which distinguishes between two persons. After that, each block stores a unique code called a "hash" that allows us to tell it from every other block.

**How the Blockchain Technology Works**

In the following steps, we explain how a Blockcahin works in detail. Let's start when a transaction is stored in a Block, and then being added to the blockchain, which consist of multiple blocks strung and linked together. In order for a block to be added to the blockchain, however, four things must happen:

i) A transaction must occur and visible. For example, when a purchase takes place through Amazon.com, it starts by clicking multiple checkout prompts by a customer, which means he/she goes against his/her best judgment and makes a purchase.

ii) The transaction is then verified when making that purchase. After that, the network of computer rushes to check the transaction from "Amazon.com" happened in the way the customer said it did. The network checks and confirms the details of the purchase, including the transaction's time, dollar amount, and participants. (More on how this happens in a second.).

iii) That transaction must be stored in a block. After the transaction has been verified as accurate, it gets the green light. The transaction dollar amount, buyer digital signature, and Amazon's digital signature are all stored in a block. Then, the transaction will likely join hundreds, or thousands, of others like it.

iv) That block must be given a hash. Not unlike an angel earning its wings, once all of a block's transactions have been verified, it must be given a unique, identifying code called a hash. The block is also given the hash of the most recent block added to the blockchain. Once hashed, the block can be added to the blockchain.

v) When that new block is added to the blockchain, it becomes publicly available for anyone to view, even the buyer, it can be viewed through Bitcoin's blockchain, all information such as access to transaction data, along with information about when 'Time', where 'Height', and by whom 'Relayed By' the block was added to the blockchain.

## Impact of Balockchain on Procurement

The Blockchain technology offers some unique benefits which can be applied to great effect in the procurement function:

i) **Increases Trust and Transparency:** The Blockchain is a transparent and safe method that requires no trust, and the procurement industry can smoothly run its transaction on it. It is through cryptography that the technology leverages to ensure digital identities of both buyer and supplier, and other parties involved are established correctly. The Blockchain can ease the way to make a contract with a supplier through a means of automating trust. By using permanently retained historical data to authenticate the supplier or distributor involved in a deal, the buyer can be assured of their trustworthiness. The technology provides transparency for both the supplier and the buyer, and thus a strong relationship can be built between them as the technology puts everything on the table for all to see.

ii) **Automated and Speed up Processes:** The Blockchain can help the procurement to speed up the process, and buyers will benefit from its efficiency by taking most time-consuming elements of a conventional transaction out of the equation, and then immediately save time and resources that would have been spent on such tasks. Using the Blockchain means sharing access databases with the supplier and other parties involved to dramatically accelerate the reconciliation process as all all of them allowed to view the same transaction.

iii) **No Middlemen:** By removing all intermediaries, it makes the processing of payments and transactions automated, and faster. This technology can also identify the nearest and most cost-effective vendors, and decreasing lead and work time, and improving business operational efficiency. Also, it helps suppliers to have access to the hiring contract.

iv) **Blockchain creates strong audit trails:** As explained the process of Blockchain, the technology stores all details and create a strong and fully documented audit trail for every transaction in the procurement, which could be used to control fraud and corruption might be committed by a buyer.

v) **Streamlined payments:** The transactions can play an important role to streamline day-to-day transaction with regard to the payment methods in the procurement. Instead of conducting lengthy processes such as involving in drafting contracts, talking to insurance and finance before finalising payment, the Blockchain technology through its smart contract system is able to dispatch payment promptly, and intermediaries (like banks and insurance) will not be necessary. A smart contract, as the term hints, is a self-executing programme that initiates a transaction only when all the preset conditions have been fulfilled. The purpose of a smart contract is to introduce some level of trust amongst transacting parties that do not know each other. Transparency exists as well since everyone can view all the details in the contract.

vi) **Better security:** The procurement department deal with hordes of sensitive data that could cost a company millions if it falls into the wrong hands. Therefore, to ensure data are safe and secure, the Blockchain technology can provide the following security methods:

> **Decentralises control:** The Blockchain breaks everything into small chunks and distributes them across the entire network of computers. It's a digital ledger of transactions that lacks a central control point. Each computer, or node, has a complete copy of the ledger, so one or two nodes going down will not result in any data loss. It effectively cuts out the middle man, and there is no need to engage a third-party to process a transaction.

> **Blockchain offers encryption and validation:** Everything that occurs in the blockchain is encrypted and it's possible to prove that data has not been altered. Because of its distributed nature, file signatures can be checked across

all the ledgers on all the nodes in the network and verify that they haven't been changed. If someone does change a record, then the signature is rendered invalid.
> **Blockchain is virtually impossible to hack:** All data recorded within nodes on the network are impossible to be hacked, and the reasons to think positively in such a way because the data is decentralised, encrypted, and cross-checked by the whole network. Once a record is on the ledger, it's almost impossible to alter or remove it without being noticed and invalidating the signature.

Restriction mode in Blockchain: Private Blockchains can be created to restrict access to specific users. Business can still benefit of a decentralised peer-to-peer network, but anyone accessing a private Blockchain must authenticate their identity to gain access privileges and it can be restricted to specific transactions.

**Advantages of Blockchain**

Here are some common advantages of the Blockchain technology:

i) **Blockchain is Immutable:** Every bitcoin network client stores the entire transaction history and are immutable, and impossible to be erased/undo the records. Each transaction is also time-stamped, transparent, and unalterable and permanent.

ii) **Zero Percentage of Fraud:** Since Blockchain is an open-source ledger, each and every transaction will be made public and hence there will be no chance of fraud taking place. The virtue of the Blockchain system will be constantly monitored by miners who keep an eye on all kinds of transactions around the clock.

iii) **No Interference:** The government or any financial institution has absolutely zero control on virtual currencies that are based on the Blockchain technology whatsoever. Hence there will be no meddling with by the governments.

iv) **Instant Transactions:** The virtual currencies/digital currencies that are based on Blockchain offer transaction times that are 10x faster than the usual bank ones, it almost takes a few minuted to be completed by the technology.

v) **Improves Efficiency and Reduces Costs:** The Blockchain technology lets individuals and companies make transactions directly to the end user without involving any 3$^{rd}$ parties. This greatly enhances the financial efficiency in every nation and lets people be less dependent on financial institutions and/or banks. Not only will save a lot of money for people in terms of fees but also other related expenses with utilising banks.

vi) **Streamlines Supply Chain Management:** As explained earlier, the Blockchain technology offers the benefit of trace ability and cost-effectiveness, and allows for the tracking of goods, their origin, quantity and more. This simplifies processes like ownership transfers, production process assurance and payments. It can also be a good means to build a good relationship with suppliers.

vii) **Security:** Each user has his/her own key to verify identity. The block encryption in the chain makes it much tougher for hackers to disrupt than traditional setups.

viii) **Traceability:** The way the information disseminated across the Blockchain makes it simple to find and solve problems efficiently, should they arise. It also creates a de facto, irreversible audit trail.

**Disadvantages of Blockchain:**

As with any technology, the Blockchain also has a few disadvantages. Here's the list of the key disadvantages of the Blockchain technology:

i) **Performance:** Such the new technology still needs more time to confirm its usability, and it remains largely untested beyond the "proof of concept" phase. And even in those tests, it has not been scaled up significantly, or run for extended periods of time to test durability.

ii) **Lack of oversight:** By definition, an open-source, decentralised ledger has no central oversight, and it could make some people nervous, and rightfully so.

iii) **Cuts out Middlemen (Jobless):** The technology could potentially save billions of dollars in intermediary bank transactions. The counter argument to this point is a lot of jobs could be redundant.

iv) **Costs:** Such costs may or may not go down with advances in technology, as the need for power and storage of data will remain a real issue, as both could cause an increase in costs and can't be avoided easily.

v) **Complexity:** The technology is not as simple as it looks like, and hence must not be handled by people who are not-so tech savvy, especially in case of currency and banking transfers, in order to avoid any crimes could targeting cryptocurrency exchanges.

vi) **Needs more resources:** The Blockchain network requires more initial resources to facilitate nodes and take care of other security measures.

## 5) Big Data in Procurement

The Big Data technology is an evolving term that describes a large volume of structured, semi-structured and unstructured data that has the potential to be mined for information and used in machine learning projects and other advanced analytics applications. In short, the Big Data is defined as any analysis activity targeted at getting more insight from large amounts of data in order to generate business value. The term big data was first used to refer to increasing data volumes in the mid-1990s. In 2001, Doug Laney, an analyst at consultancy Meta Group Inc., expanded the notion of big data to include what is know "3Vs"; Volume, Velocity and Variety. Later, two more Vs were added; Value and Variability. Thus, the big data technology is currently charachtrised as 5Vs:

i) Extreme *volume* of data.
ii) The wide *variety* of data types generated by an organisation.
iii) The *velocity* at which the data must be processed.
iv) The *value* of data.
v) The *Variability* of data.

The technology encompasses a wide variety of data types such as:

i) Structured data in SQL databases and data warehouses,
ii) Unstructured data, such as text and document files held in Hadoop clusters, or NoSQL systems.
iii) Semi-structured data, such as web server logs or streaming data from sensors.

The most important part in the Big Data technology is the data analytics techniques, which provides a means to analyse data sets and draw conclusions about them to help organisations make informed business decisions. Business Intelligence (BI) queries, answer basic questions about business operations and performance. Further, the analytical techniques involve complex applications with elements such as predictive models, statistical algorithms and what-if analysis powered by high-performance analytics systems, which helps in improving business in terms of new revenue opportunities, more effective marketing, better customer service, improved operational efficiency and competitive advantages over rivals.

**Types of Big Data Analytics**

The following are the main types of analytics and the scenarios under which hey are normally employed:

i) **Descriptive Analytics:** As the name implies, this type is used for descriptive analysis or statistics, which can summarise raw data and convert it into a form that can be easily understood by humans. They can describe in detail about an event that has occurred in the past. This type of analytics is helpful in deriving any pattern of interpretations for any events in the past events, for framing better strategies for the future. This type is the most frequently used types of analytics across organisations. It's crucial in revealing the key metrics and measures within any business.

ii) **Diagnostic Analytics:** This type is considered as a successor to descriptive analytics. It is used as a diagnostic analytical tool to dig deeper into an issue at hand so that they can arrive at the source of a problem. In a structured business environment, tools for both descriptive and diagnostic analytics go hand-in-hand.

iii) **Predictive Analytics:** This type is used to predict and forecast the probability of an event may be happening in future, along with estimating the accurate time for its occurrence. The model relies on different variables to predict the future. For example, in the healthcare domain, prospective health risks can be predicted based on an individual's habits/diet/genetic composition. Therefore, these models are very important across various fields.

iv) **Prescriptive Analytics:** This type explains the step-by-step process in a certain situation. For instance, a prescriptive analysis is what comes into play when the Uber driver approaches the easier route from G-maps. The best route is chosen by considering the distance of every available route from the pick-up route to the destination and the traffic constraints on each road.

v) **Cyber Analytics:** This type is a combination of cyber security skills and analytical knowledge. The model is a new and rising proficiency within the business and the data analytics to face any cybersecurity threats which have been escalated in volume and sophistication in recent years. Cyber analysts use sophisticated tools and software to pinpoint vulnerabilities and close off attack vectors using a data-driven approach.

## Impact of the Big Data Technology on Procurement

The following are the key impacts of the Big Data technology on the role of procurement, and it is called as "procurement's secret weapons for the next couple of years" by many professionals.

i) **Analytical power:** The technology is used to analyse datasets used by the procurement function. The five key aspects in procurement are enough to be analytically assessed to ensure having the right data for better strategic plans and decision. The five aspects are; spend analytics, category analytics, supplier analytics, compliance analytics and performance analytics. The analytics field of the Big Data is also able to process complex data sets from a broad range of ERP systems, consolidating relevant information from multiple streams onto single dashboards in order to inform business decisions. With the help of data, procurement teams can become a more strategic advisor to steer the direction of the business.

ii) **Combine disparate datasets:** The technology is used to combine datasets from different sources such as contract data, spend data, project portfolio management, performance management, talent management, and supplier performance. Therefore, the power of Big Data can lead to greater opportunities for companies to predict market trends, spending, consumer behaviour and procurement needs in real-time. By combining all data, Management can move away from making reactive decisions on fiscal data and instead take a forward-looking approach that is insight-led.

iii) **Run more queries, faster:** The Big Data enables companies to run queries much faster. For some, this might involve running more and more scenarios to optimise sourcing award decisions based on internal constraints and different market forecasts. In other cases, the ability to instantly pivot on a massive spends cube (or create and toss out new cubes after quickly drilling for new opportunities) and look at spend in new ways (by different locations, taxonomies, material codes, etc.) will become the norm in leveraging speed that comes with new Big Data approaches.

iv) **Solve complicated challenges:** The Big Data allows the procurement to not only make better award decisions with an eye toward total cost, but to optimise for lower cost structures before sourcing events themselves (e.g., by suggesting changes to specifications and tolerances, where possible, enabling and starting collaborations between design/engineering, business owners, and procurement teams).

v) **Better Management:** The technology allows better implementation, management, measurement and forecast, savings, and cost avoidance. The Big Data is able to combine, report and analyse multiple datasets leveraging internal cross-systems data combined with external market information to track and manage savings will have a transformative effect on organizations.

vi) **Gain a predictive edge:** The Big Data is able to introduce a new scenario planning, predictive modeling and forecasting competencies into procurement. And those who do it right will truly gain a new predictive edge in the market. For example, Analytics help an organisation to achieve the competitive edge through tie together an internal data sources with external data sources, such as supplier databases.

vii) **Improves forecasting /decision making:** The Big data can provide the ability of decision making based on real evidence. For instance, by collecting and analysing suppliers' past performance data, as well as current market pricing and risk assessments, sourcing teams can take a data-driven approach to awarding contracts to suppliers, rather than awarding them based on lowest price, an existing relationship, or "gut feel."

viii) **Profit Generation:** The at Big Data presents significant opportunities and can transform an organisation from one that simply manages costs to one that generates profit. Data already collected, often insufficiently analysed, can be used straight away to increase the value that sourcing professionals bring to the organisation and increase the role professionals will play as business practices evolve.

ix) **Manages supply risk:** A powerful application of Big Data lies in identifying trends and events that act as warning signs for supply-chain risks. For example, the technology can help continuously update about suppliers' and sourcing markets' risk profiles and even trigger contingency plans in case of, e.g., bankruptcies or natural disasters.

x) **Replaces Human's brain:** The big data could be a replacement to management's mentality as they might be considered as a hindrance in taking the right decisions, especially with regard to some complected qualitative methods.

xi) **Sourcing cost improvements:** Big Data analytics is a great source from which to identify opportunities for better sourcing, and can also discover solutions and alert the organisation to act, e.g. to renegotiate contracts as soon as there is a significant decrease in the price of crude oil.

xii) **Organisational efficiency and agility:** The Big Data solutions link and aggregate all relevant information, thereby facilitating and speeding up strategic and operational procurement activities significantly.

**Advantages of the Big Data Technology**

The following are multiple advantages of the Big Data technology that be in benefit of organisations:

i) It provides organisations for better decision-making and insights towards competing and growing.
ii) It increases business user productivity through the system's analytical power. e.g. it could analyse more data, more quickly, which increases their personal productivity, and thus allows an organisation to increase productivity more broadly.

iii) It helps to increase operational efficiency and reduces costs through decreasing expenses, which is one of the Big Data objectives.
iv) It improves customer service through having a wealth of information about customers, which helps to provide better services to meet their demands.
v) It helps companies to detect fraud by using the Big Data analytics to look for patterns of fraudulent behavior in enormous amounts of unstructured and structured data, and it can be done in real time.
vi) It helps to increase revenue as the technology can play important roles in terms of improving the decision making method, better customer services, having accurate information about the market status, and preventing any fraud.
vii) It helps to increase business and IT agility through faster and more frequent changes to their business strategies and tactics.
viii) It provides a greater innovation and initiative to lead the market through a glean insights that their competitors don't have, they may be able to get out ahead of the rest of the market with new products and services.

## Disadvantages of the Big Data Technology

The following are some disadvantages of the Big Data that need to be considered by a business:

i) Need for talented and skilled employees to handle the technology, and data scientists and big data experts must be considered as the most highly coveted.
ii) Data quality and accuracy must be available before proceeding any kind of analysis. Otherwise, the results may be worthless, or even harmful if acted upon.
iii) Need for cultural change in an organisation and any resistance to change could lead to bad results.

iv) Information collected for big analytics efforts must be complied with government regulations and industry standard, as some data are sensitive, especially what are related to persons.
v) Sensitive data are prone to cybersecurity risks if not protected well.
vi) The Big Data needs for a storage space to house the data, networking bandwidth to transfer it to and from analytics systems, and compute resources to perform those analytics, which are all expensive to purchase and maintain. Additionally, companies may face significant expenses related to staffing, hardware, maintenance and related services, which could run significantly over the allocated budget.

## 6) Negotiation via Videoconferencing

As explained earlier, Negotiation is an important part in the procurement process, which covers the period from when the first communication is made between the buyer and the supplier through to the final signing of the contract, and such the process could take a bit long time. However, the negotiation process is costly, as buyers or purchasing negotiators need to leave his/her office and travel for a long distance. With new technologies, especially with the maturity of video conferencing such as Skype or Google Hangout to interact in real time, organisations now consider video conferencing as a preferred tool for communicating with suppliers.

### Benefits of Negotiating via Videoconference

i) Video conferencing is relatively cheap to use, it only requires an application either free or premium apps such as e.g. Skype for business, Google Hangout, Join.me, RingCentral Meetings, and Cisco WebEx., etc. Such apps can either be used through the computer desktop or Smartphone. After that a buyer just needs an agreement with the supplier to link up at a certain time and date.

ii) Video conferencing is perceived as a 'rich' communication medium that allows for win-win bargaining and mutual gain, because it allows people to learn from each other's visual and verbal cues.
iii) Video conferencing allows parties to communicate both verbally and nonverbally, through enabling them to jointly view and discuss documents, slide shows, and videos.
iv) No need to think about travelling costs, video conferencing lets a buyer to negotiate from her/ his office.

## The Limitations of Negotiating via Videoconference

i) Videoconferencing means limited visibility. It means that negotiators can only see their heads and upper torsos showing on the screen.
ii) Weak Internet connection or poor technology can result in a grainy or choppy image that makes it difficult to read a counterpart's facial expression. Background noise or a busy background may also cause distractions.
iii) Impossible for negotiators to truly make eye contact during a videoconference, since computer cameras tend to be located at the top of the screen, when we stare at our screen, we appear to be looking downward rather than into our counterpart's eyes. This lack of eye contact might impair negotiators from building trust and rapport.
iv) Technical difficulties are possible and may crop up during the conference. Such technical problems could interrupt the flow of a negotiation or leave negotiators feeling irritated.
v) Privacy and security challenges are another issue with video conferencing. When the privacy of a negotiation is paramount, videoconferencing may pose special concerns, such as secretly recording the conference. Thus, when security is critical and trust is low, the negotiator may require to make an extra effort to negotiate in person.

# CHAPTER 9

# Technology Applications in Distribution

## 1) Introduction

Modern technology has also significantly changed the distribution trends as the third critical element within Material Management, according to the book structure. In this chapter, we will discuss all new technologies that could be used in the distribution management for advancing strategy processes in Material Management, as the only way to gain the competitive edge in today's fast-paced global markets. The chapter covers the role of new technologies and information systems in distribution management, especially the physical distribution part, which is considered the key fraction of marketing and are now directly linked to sales in some companies. Therefore, improving physical distribution aspects means ensuring the best service could be provided to customers. Moreover, new innovations in the distribution sector have a critical role in the evolution of the world economy, and some companies have already started thinking seriously of the applications of technology to conquer new markets through strengthening distribution to win new consumers with the convenience that modern technology can provide.

Some common modern technologies between and among distribution, warehousing, and procurement have already covered in the previous chapters, for that reason the chapter is structured by mere focusing on the distinguished technologies and innovations that are only applicable and developed for the distribution management aspects.

**Main Technologies offered to Distribution Management**

## 2) Control Tower Solution

The "Control Tower" technology is considered a comprehensive solution through enabling control of end-to-end supply chain performance across Plan, Source, Make, and Deliver functions. The technology is defined by Capgemini Consulting, Global Supply Chain Control towers, as *"a central hub with the required technology, organization, and processes to capture and use supply chain data to provide enhanced visibility for short and long term decision making that aligns with strategic objectives"* (Capgemini 2017). The control tower is implemented either by outsourcing companies through using a 3PL, 4PL or LLP (Lead Logistics Partner), or by developing an in-house capability. The term "Control Tower" merely implies the objectives of the technology, which is developed to control information flows from two different ways:

i) **Control Information:** Making it possible to control processes.
ii) **Accountability information:** Using it for insights about the processes and flows within the supply chain, using the Control Tower KPI dashboard.

The above objectives are classified as mainstays in improving decision making and business performance within the supply chain processes.

The technology that is applied to the Control Tower through an intelligent software package to make interchains within the whole cycle from one centralised point, which is considered as the director for all communications within the supply chain. Therefore, it would be so much easier to keep an overview of all logistics activities, performances and distributions. By combining all the logistics data with the Control Tower tools it will be possible to:

i) Gain visibility into business logistics processes.
ii) Prepare scenarios for the future.
iii) Invent solutions to avoid issues.
iv) Establish partnerships within supply chains.

Moreover, the technology can be used for better planning, proactive event management, improvement of the performance of supply chain partners, and more sophisticated supply chain analytics.

**Components of Control Tower**

There are four key components of the Control Tower to make it work effectively, and missing one of them will not let the system work effectively:

a) **Processes:** Using the high maturity of the process to support the logistics and supply chain activities statistics
b) **Technology:** Different systems used in the Control Tower for for tracking, planning, and data to meet the following three business level:

  i) Operational: for real-time information.
  ii) Tactical: for efficient and effective.
  iii) Strategical: for opportunities exploration and optimisation.

c) **People:** Who with Hard & Soft Skills run the Control Tower. Their main duties are:

  i) Planner (daily execution).
  ii) Logistics Engineers & Data analysts.
  iii) Customer Service.

d) **Knowledge:** It is considered as an extra pillar to add value to the scope of the Control Tower, in order to enable the system perform the activities at a higher level.

## Key Benefits of the Control Tower in Distribution

The following are the key benefits of the Control Tower for distribution:

a) **Visibility**: This is one of the most important aspect of the Control Tower. The Control Tower provides global visibility to all stakeholders, including, a shipment's location and expected time the items Arrive (ETA). Beside, the technology provides visibility about the data concerning freight, performance, and costs of the shipment. Thus, such the visibility to data enables businesses to set up appropriate strategic planning of upcoming shipments, improved performance and analysis of shipments.

b) **Performance Improvement**: The technology can help businesses to improve their processes through linking up globally based on a standard global process. Therefore, it would be a great opportunity for them to follow a customary procedure which could bring a greater supervision over the processes while also lowering lead times and costs, and hence achieve an improved performance.

c) **Savings**: One of the key role of the Control Tower is to look for opportunities about shifting transport modes, changing service level and consolidating freight. Therefore, the technology can drive a huge savings while maintaining the highest level of performance of the carriers.

d) **Metrics & Analytics**: Through applying the technology in the distribution management, companies are able to manage a successful process through using the KPI measurement tool. The following are some metrics available via the Control Tower for best distribution measurement:

i) Volume by mode, service level and lane.
ii) Carrier Performance: on-time pickup and delivery.
iii) Cost: Per milt/kg, mile/km.
iv) Lead time by lane.

e) **Technology**: The technological aspect of the Control Tower solution provides many advantages such as; real-time decision-making based on an end-to-end (e2e) view of the supply network. Also, it provides flexibility, multi-leg shipments, equipment type, auto tender, rating engine, automated premium approvals and a smartphone application, and etc. The technology involved allows for better managing in lower costs.

**Control Tower' Strategy in Distribribution Management**

From the distribution's point of view, improving distribution management is considered the key aspect of the Control Tower technology, and thought it could be used exclusively for distribution management for making centralised planning and execution of localised transportation functions. Furthermore, the technology is defined by some practitioners and scholars as a service provided by transport companies in order to provide industry specific solutions by directing transports through a network, and to gain control of the information flow for transportation, inventory and order activity, and managing those activities from a single location using an online web portal, which is hosted in the cloud, logistics information is always and everywhere accessible. Furthermore, one of the roles of the Control Tower from the distribution aspects is to monitor the operations through a network that is dependent on several factors such as border crossing, consignment collections and deliveries in several regions and the use of multiple transportation modes and merge-in-transit. There are two types of Control Tower implemented in practices to enhance the distribution management by organisations:

a) **Lead Logistics Provider (LLP) Logistics Control Tower's activities:** This type is about outsourcing Control Tower activities operationally and tactically to a logistic service provider (Third party), to carry out the tasks based on the 3PL or 4PL concept, such as:

i) Transport Planning & Monitoring.
ii) Tender Management.
iii) Warehouse Management.
iv) Order Management.
v) Network Design.
vi) Exception Handling.
vii) The Management and Control of all Related Information.

DHL is a good example in this regard. The company uses the Control Towers for managing the premium freight service utilising a DHL's extensive network of expediting carriers operating in all markets. The LLPs operate a controlled premium freight solution through a range of approved carriers, ranging from express van and motorcycle deliveries to aircraft charter, assuring has sufficient resources available at all times to ensure an uninterrupted service and at a controlled cost.

    b) **In-house Control Tower's activities:** This type means the company owns full control of Control Tower activities, and instigate its business based on own function strategy. In addition to control activities operationally and tactically, it includes controlling the activities strategically by the company. The following are the main activities of this type:

       i) Procurement.
       ii) Sourcing.
       iii) Sales and Operations Planning (S&OP).
       iv) Order Management.
       v) Direction of Logistics Activities.
       vi) The Management and Control of all Related Information.

## Tasks of the Control Tower in Distribution

The following are the main tasks being carried out through the Control Tower:

i) Receiving pick-up orders.
ii) Transport planning in cooperation with transport companies involved.
iii) Forwarding transport requests to transport companies and supervision of all shipments the whole way.
iv) Problem solving in case of exceptions with transport companies and the customer.
v) Coordinating trailers to the right loading/unloading places according to the customer's priority list.
vi) Informing the customer about operations at all stages.
vii) Assuring to fulfill the set KPI's and their reporting requirement.
viii) Making visibility over the order, drivers and third party logistics partners.
ix) Having a real-time tracking, which can be done through smart devices to create an easy communication with drivers as well as allows having visibility and managing orders easily.
x) Providing omni-channel access by having access to information on any device and be able to share it and collaborate in real-time.
xi) Providing data analytics to efficiently direct the field workforce and last-mile delivery to their intended destinations, speeding up the delivery process while maintaining the quality of the delivery process.
xii) Providing notifications and alerts to resolve disruptions before they disrupt the business.

## The Control Tower role for enhancing distribution

The following can be considered the key role for using the Control Tower for enhancing the distribution management:

i) Improvement in vehicle utilisation and therefore reduction in fleet size.
ii) Enables to exchange information with all clients at each stage along the supply chain.
iii) Proactive selling of the unused space (multi-customer consolidation, backhauls).
iv) Reduction in miles travelled to lower costs and support the company's green policies.
v) Enables the logistics organisation to act as a lead provider for many customers in a consistent manner.
vi) Offers the customer the benefit of being part of a wider network to manage unpredictable demand, such as from e-Fulfillment channels.
vii) Implementation of a platform for further innovation that can be easily deployed to all customers.
viii) Helps to gain insight over carrier performance to evaluate trends and pursue advantageous alternatives.
ix) Ensures the most appropriate distribution option at the lowest cost.
x) Reacts in real-time to deviations from planned transportation movements.

## Drawbacks of the Control Power

There are some limitations as well in the Control Tower as follows:

i) **Limited visibility:** Control Tower vendors may claim to offer end-to-end visibility, but the truth is that they usually offer a simplistic form of visibility into one part of the supply chain at the expense of another. In other words, critical information that affects business' profitability may reside with the logistics providers or in retail stores, which can affect to take the right decisions without having full and clear information.

ii) **Latency of Response:** The latency between detecting a discrepancy and translating its effect on the strategic, operating, and execution plans, and then finding the root cause and taking corrective action, can vary from one week to one month. Furthermore, even these sluggish results require a highly manual process where human planners are brought in to determine the optimal response.

iii) **Effortful Job:** No execution, almost every company today has a separate system for planning, execution, and business intelligence (reporting and alerts). Multiple and asynchronous systems require large numbers of people to sift through vast amounts of data, detect business problems, and take corrective actions. The Control Tower approach does nothing to change this state of affairs. At best, the Control Tower can enable business to manually adjust plans more quickly, but even in the highly unlikely event that it re-plans for business, and new plans are of little benefit until they being executed.

iv) **Not scalable:** The Control Tower solutions will struggle to scale across the entire business, especially if the goal is to manage at the SKU/item level of detail. One reason is the heavy emphasis that control towers placed on human planners. However, no matter how people intelligent, but they cannot keep up with the millions of data points that today's global supply chains are generating. The lack of scalability also becomes a problem when trying to build connections with trading partners, which could lead to a flaw.

v) **Trading partner connections aren't reusable**: At best, the most advanced versions of Control Towers are using the "hub-and-spoke" model to connect trading partners to each other. This works by connecting a single company (hub) to surrounding companies (spokes). The major drawback with this approach is these connections aren't reusable. Thus, if a spoke wants to connect to other spokes, a whole new set of connections must be formed, which is unlikely given the amount of effort it takes to form a connection in the first place (integrating systems, establishing new business processes, etc).

## 3) Intelligent Transportation System - ITS

ITSs are technology implemented in logistics operations for creating new opportunities for the efficient movement of goods. Such the technology is important for cross-linked itineraries across multiple modes of transportation to smooth the distribution operations as a systemic solution to delays, and to improve the exchange of information and real-time status updates regarding different business operations in different modes of transportation. Moreover, the ITS is implemented to improve economic performance, safety, mobility and environmental sustainability. The term intelligent transportation system (ITS) and smart transportation management (STM) systems are used interchangeably in the literature to refer to the identical concept for using advanced information and communication systems for managing and controlling transportation operations, and it is used in various service and business concepts. For example, the system is used by public transportation to ease up the control and enhance the surveillance of the buses in a fleet of buses, and to enhance performance, reduce accidents, optimise fuel consumption and enable multimodal transport through integration of Information and Communications Technology (ICT). It is also used by drivers in their cars for assisting them throughout their driving. However, our interest is only the usability of the ITS technology from the physical distribution aspect, and how such the technology can enhance operations in logistics.

**The Usability of ITS**

The following are the main usability of the ITS technology:

    i) Routing.
    ii) Traffic management and congestion avoidance.

iii) Traffic signal coordination.
iv) Truck monitoring.
v) Parking space management.
vi) Safety information and warning announcements.
vii) Weather information.

**The importance of the ITS technology**

i) Makes transportation system more efficient, secure, and safe, through the use of information, communications and control technologies.
ii) Improves the attractiveness of transport in business and public.
iii) Tackles rising congestion, which increases travel times and industry costs.
iv) Reduces the environmental impacts of transport.

**Characteristics of ITS**

The following are the main characteristics of ITS in today's business:

i) The ITS technology is based on data collection, analysis and using the results of the analysis in the operations, control and research concepts for traffic management where location plays an important role.
ii) The technology is the integration of a global structure facilitating the cooperation between several systems of transportation previously isolated, such as electronic imaging, data processors, information and communication systems, roadside messages, GPS updates and automated traffic prioritisation signals etc.
iii) ITS combine cutting-edge technology such as electronic control and communications with means and facilities of transportation.
iv) The system involves vehicles, drivers, passengers, road operators, and managers all interacting with each other and

the environment, and linking with the complex infrastructure systems to improve the safety and capacity of road systems.
v) ITS improves transportation safety and mobility and enhances global connectivity by means of productivity improvements achieved through the integration of advanced communications technologies into the transportation infrastructure and in vehicles.
vi) The ITS encompass a broad range of wireless and wire line communication based information and electronics technologies to better manage traffic and maximise the utilisation of the existing transportation infrastructure.
vii) The technology helps for advanced planning and forecasting systems needs to provide timely and accurate information to managers and thus, give useful insight and perspectives, which can be considered an opportunity of improving the supply chain reliability and predictability and thus, having more visibility to better information and clearer views of the entire supply chain.
viii) ITS relying on wireless vehicular networks has the potential to become the platform that overcomes problems related to technology proliferation like reliability, connectivity, limited range, scalability and security.

**The Impact of ITS on Physical Distribution**

The ITS technology plays an important role in the logistics distribution system, especially in freight transportations. The following are the impact of ITS on physical distribution:

i) It helps towards co-modality by improving infrastructure, traffic and fleet management and facilitating a better tracking and tracing of goods across the transport networks.
ii) The ITS technology can be used for making an e-Routs approach whereby information on the location and condition of transported goods can be available online in a secure way.

iii) The ITS is able to work intelligently with the cargo activities through making goods become self, context and location-aware as well as connected to a wide range of information services.

iv) Having implemented the ITS, all information associated to the physical flow of goods will become a paper-free, and it will develop the vision of a paperless information flows accompanying the physical shipment of goods. Especially when implementing it in e-Freight.

v) The ITS technology can make a tremendous change in the supply chain by incorporating the chains based on navigation systems, digital tachographs and tolling systems.

vi) The ITS technology helps to make traffic management more efficient by promoting intelligent transport systems as well as facilitate the rollout of innovative services.

vii) The ITS technology allows the exchange and coordination of information between vehicles and the road infrastructure, as well as the exchange of information with third party logistics. In other words, the technology helps to optimise freight processes inside cities, through the appropriate exchange of information between and among all key stakeholders during the transportation process.

viii) The ITS helps to improve customer service as well as to obtain a cost reduction through intelligent transport systems and their interaction with transportation management systems (TMS).

**Advantages of the ITS technology**

The following are the summary of the main advantages of the ITS technology:

i) Reduction of operational costs of the toll collection system.
ii) Increase the traffic capacity.
iii) Reduces time for drivers.
iv) Reduction in traffic congestion.

- v) Reduces service time.
- vi) Increases traffic safety.
- vii) Adds opportunities for logistics via e-Commerce.
- viii) Reduces fuel consumption.
- ix) Reduces carbon dioxide and other substance emissions.
- x) Increases user/customer satisfaction.

**Disadvantages of the ITS Technology**

The following are the negative side of the ITS technology:

- i) Implementing such the technology is considered as capital intensive either for business or government.
- ii) Such the technology requires similar monitoring, information dissemination and traffic management systems on the adjacent arterial road systems to achieve real benefits.
- iii) The implementation of the ITS elements at toll roads entails toll payment deficiency.

## 4) Global Positioning System - GPS

GPS tracking application development has changed the idea of transportation and logistics business and its emerging could help the business in numerous ways. The GPS was first designed and implemented by the United States Department of Defense; and was called NAVSTAR (Navigation System with Timing and Ranging). The system was first launched in 1978 and by the mid 1990s there were 24 satellites, all of which are still in operation today. The GPS technology is used in various tasks, but it is mainly an important means of today's business, especially in the fleet management. The technology is implemented in distribution logistics to provide an essential service to industrialised societies by transporting finished goods and raw materials over land, typically to and from manufacturing plants, retail and distribution centers. Furthermore, the GPS technology is developed based on the mobile communication using the integration of wireless technologies whereby vehicles, assets, and staff over a wide geographic area in a real time for the sake of improvement, the accuracy, and efficiency of logistics management. Generally speaking, the GPS is mainly used to manage all aspects of fleet transport and logistics operations, which is known as the fleet management.

**Automatic Vehicle Location AVL based on GPS**

The GPS is the key tool for AVL to locate vehicles in real time, and such the application is essential to know the precise location of a vehicle. Also, the application is used to deal with criminal activity, medical emergencies, or mechanical breakdowns. The application is currently used by many sectors as an important feature for monitoring a vehicle, management, and thus helps management make decisions about their operations, increase efficiency, and cut operating costs.

**Basic Modules of GPS**

i) **Digital Map Data Base Module:** A digital map database is a dispensable module for any vehicle and navigation system that involves mapping-related functions. Without a map, it is very difficult for a traveler to explore an unfamiliar area and make correct decisions concerning the route. With a map as a medium, complex information can be communicated very easily.

ii) **Positioning Module:** Positioning involves determination of the coordinates of a vehicle on the surface of the earth. Three positioning technologies are most commonly used; stand alone, satellite based, and terrestrial radio based. Dead reckoning is a typical stand-alone technology. A common satellite-based technology involves equipping a vehicle with a global positioning system (GPS) receiver. Dead reckoning and GPS technologies have been used widely in vehicles.

iii) **Map-Matching Module:** To provide drivers with proper maneuvering instructions or to correct display the vehicle on a map in an error-free fashion, the vehicle location must be precisely known. Therefore, an accurate vehicle location is considered a prerequisite for good system performance. Dead reckoning can track the position of a vehicle relative to another position, such as an origin. Typically a vehicle heading and distance traversed are used to determine incremental changes in the position of the vehicle relative to the origin. When the dead-reckoning behavior indicates that vehicle is in a certain position on the map, the vehicle position may be adjusted to some absolute position on the map. This will eliminate the cumulative error until the next map-matching step.

iv) **Route-Planning Module:** Route planning is a process that helps vehicle drivers plan a route prior to or during a journey. It is widely recognised as a fundamental issue in the field of vehicle navigation. Route Planning can be further classified into either multivehicle (system-wide) route planning, which plans multidestination routes for all vehicles on a particular road network, or single-vehicle route planning, which plans a single route for a single vehicle according to the current location and a given destination.

v) **Route Guidance Module:** Guidance, an integral part of Intelligent Transport System (ITS), is the process of guiding the driver along the route generated by the route planning module. Guidance can be given either before the trip or in real time while on route. The pre-trip guidance could be presented to a driver as a printout. These instructions might include turning, street names, travel distances, and landmarks. On the other hand, en-route guidance would require providing turn-by-turn driving instructions to a driver in real time. It is much more useful, but requires a navigable map database, an accurate positioning module, and demanding real-time software.

vi) **Human-Machine Interface Module:** Human-machine interface is a module that provides the user with the means to interact with the location and navigation computer and devices. To develop a successful human-machine interface, a certain procedure must be followed that may include identification of requirements, determination of functions to be supported, specification of interface type(s), selection of controls and displays, and finally, designing and implementing these interfaces.

vii) **Wireless Communication Module:** Wireless data applications in Intelligent Transport System (ITS), plays a critical role in making the vision of mobile computing a reality. It provides a very valuable opportunity to present relevant information to the vehicle and its occupants as well as to obtain data for transportation management systems. Many quality services can be provided to drivers using communication technology.

**Types of GPS Technology**

There are different types of GPS trackers used to monitor both vehicles, goods, and people. The following are the main ones offered to businesses and individuals:

i) **Personal Trackers:** This type is used to monitor people through a personal device like a pocket chip or bracelet, and it is useful to constantly follow and locate people, animals, or any other objects.

ii) **Asset Trackers**: This type is similar to the previous one, but used for non-vehicular items. Asset trackers can be anything from a small radio chip to large satellite tags. This type is mainly used by warehouses and supermarkets in order to prevent theft.

iii) **Cell-based GPS vehicle tracking:** This type is one of the most common types of GPS tracking, which is used by businesses using either cellular or satellite networks. This type uses a device to capture data from the vehicle and then reports the data by using cellular towers. Compared to satellite tracking, cell-based vehicle tracking costs less and reports faster. This type is mainly used by companies to monitor deliveries in order to simplify their customer service workflow.

iv) **Satellite-based GPS vehicle tracking:** This type works through satellite networks and businesses can get updates from even the most remote locations. This type is able to provide constant updates and drivers can use it instead of their cell phones.

v) **Cellular-based tracking:** This type uses a phone app via the Internet data to record and report its location. This type is able to track employees, contractors, and others even without installing an in-vehicle tracking device.

**Advantages of GPS**

The following are the main advantages of GPS:

i) **Reduce operating expenses:** The technology allows to choose the best and shortest routes for vehicles, which in turn helps the business to reduce fuel consumption, unnecessary overtime costs.

ii) **Save time:** The technology helps to save time through guiding the drivers to avoid the busy streets in real-time basis. Without the GPS tracks could be stuck in case of any traffic jams.

iii) **Reduce downtime:** The technology helps to make informed decisions and schedule trips more efficiently, thus reducing the downtime of vehicles. When reducing downtime, business is able to improving the productivity and profitability of a transportation company.

iv) **Optimise resources:** The technology allows businesses to monitor vehicles and get detailed insights on fuel usage, driver behavior, engine idling, etc. in real time. Thus, management is able to utilise resources more effectively and hence identify key money-saving areas.

v) **Reduce insurance costs:** The technology helps businesses get some premium from insurance companies who normally provide special discounts on vehicles that are equipped with GPS tracking systems.

vi) **Maximise vehicle utilisation:** The technology allows the most out of transportation vehicles as the GPS tracking software eliminates vehicle idle time, reduce over-speeding and decrease fuel consumption.

vii) **Be more predictable:** The technology is able to provide customers with accurate delivery times, real-time tracking info and other minute details.

viii) **Manage the field staff and drivers effectively:** The technology helps the fleet managers in managing the drivers effectively through pulling up the data obtained from the GPS tracking system, analyse it and take corrective actions whenever necessary.

ix) **Optimise financial management:** The technology allows businesses to analyse the routes taken by drivers, which in turn helps them to make a comprehensive profit and loss analysis and hence allocate the budget accordingly.

x) **Increase the number of trips:** The technology helps to make better plans through assigning tasks whenever they spot an idle vehicle. Thus, businesses are able to increase the number of trips per day. More trips translate to more business.

xi) **Locate assets:** The technology helps to locate any lost or stolen vehicles/goods could be happening. With the advanced features of the GPS, the fleet manager is able to locate the vehicle and of course goods being shipped within a real time base.

xii) **Reduce maintenance costs:** With the advanced features of GPS equipment, such as vehicle diagnostics, fuel level indicators, engine temperature indicator, the fleet manager is able to monitor key aspects such as engine oil and the overall health of vehicles. Therefore, the business can save a lot of money for a routine inspection.

xiii) **Ease of use:** The technology is very friendly, where the critical information such as location data, vehicle status, driver's behavior can be accessed easily with the help of a GPS fleet management software via mobiles, tablets and laptops.

xiv) **Get timely alerts:** The technology has many features such as SMS and email alerts. Therefore, it can be used when any vehicle goes out of the business zone. Besides, the software can send timely notifications in case of accidents and other mishaps so that the fleet manager can take the necessary steps.

xv) **Improve Safety:** The technology helps to improve the safety through monitoring the fleet in real-time. In case of untoward incidents, the fleet manager can provide direct assistance and support to drivers. The GPS fleet management systems help implementing two-way communication between drivers and fleet managers.

xvi) **Be competitive**: The technology is considered an important tool to stay strong in the market through streamlining the transportation aspects. The GPS is able to reduce costs, provide first-class customer service and keep employees happy as well.

## The Impact of GPS on Logistics Distribution

i) The GPS technology is considered more prevalent in contemporary and sophisticated physical distribution which include multi-tiered suppliers, and manufacturers that are globally dispersed.

ii) The GPS can enhance the efficiency of deliveries to the customers by navigating and routing of deliveries along the supply chain.

iii) The GPS is considered as distribution solutions through monitoring the life cycle of inventories from purchase-production-sales.

iv) The GPS can provide managers more accurate estimates of both the time of arrival and the time of delivery of goods to the customer, which helps to achieve customers' requirements on time.

v) The GPS helps the fleet manager to control operations practically through managing a vehicle fleet to improve scheduling, operating efficiency, and effectiveness.

vi) The GPS helps the fleet manager to supervising the use and maintenance of vehicles and associated administrative functions, including coordination and dissemination of tasks and related information to solve heterogeneous scheduling and vehicle routing problems.
vii) The GPS helps the fleet management with a consequent saving of fuel and time.
viii) The GPS helps the fleet manager to ensure safeguards against shipping and freight fraud, and prevent any kind of scams.
ix) The GPS can provide 24-hour, 3-dimensional position, velocity and time information to the fleet management anywhere on the surface of the Earth.

**Disadvantages of GPS**

The following are some disadvantages of GPS Tracking System:

i) Sometimes the GPS device may fail due to certain reasons, so that carrying a backup map and directions might be required.
ii) The GPS device uses a battery and in case of any failure, it could delay some important jobs.
iii) The GPS signals are not accurate due to some obstacles to the signals such as buildings, trees and sometimes by extreme atmospheric conditions such as geomagnetic storms.
iv) Sometimes due to poor network and technological errors, the GPS might show wrong position and route, which misdirects the users.

## 5) Geographical Information System - GIS

A GIS is a collection of computer hardware, software, and geographic data for capturing, managing, analysing, and displaying all forms of geographically referenced information. Choosing sites, targeting market segments, planning distribution networks, responding to emergencies are all problems that involve questions of geography and are all relevant to businesses and the government. The GIS technology is used for many purposes, such as to determine optimum locations for roads, railroads, airports, utilities, subdivisions, retail market outlets, and hazardous waste facilities. In all purposes, businesses or governments use the GIS to view, understand, question, interpret, and visualise data in many ways that reveal relationships, patterns, and trends in the form of maps, globes, reports, and charts. The GIS technology, which was proposed by Roger Tomlinson in 1968, is considered now an important element to gain a competitive advantage as it concerns with the decision level that could improve operations in business, as it could meet the business requirements in terms of transaction processing, operations, inventory control, planning and decision-making, and internal management and control. Furthermore, GIS is a powerful tool in market analyses because it also provides a way to bring together data from multiple sources and link them based on spatial attributes. This often involves a process of layering different types of data on the same map projection so that the decision maker can identify and visualise how data intersect and interact. Thus, GIS is a useful and unique query tool for accessing and displaying components of a database based on the data's spatial characteristics. In brief, the GIS serves businesses through four ways:

i) **Create** geographic data
ii) **Manage** it
iii) **Analyze** it
iv) **Display** it on a map

From above the above ideas, we can understand that the GIS' main objectives are the following:

i) **Data Management:** The GIS is used to input and store data, to retrieve that data through spatial and conditional queries, and to display the results. For the data management type of application, the GIS is merely used as an inventory system with the purpose of storing and displaying information about the spatial features. These features are things like the width, number of lanes, and traffic count for a particular highway.

ii) **Analysis:** Then, the GIS system analysis the data in order to determine the best solution. For example, in case of distribution, determining the shortest path between two locations, grouping of areas of land into larger ones depending on certain criteria, and so on.

iii) **Decision Making:** Management then can take the right decision based on the results obtained from the analysis level.

**Component of GIS**

i) **Hardware:** Hardware is the physical component of the computer and GIS runs on it. Hardware may be a hard disk, processor, motherboard and so on. All these hardware work together to function as a computer. GIS software run on this hardware. The computer can be standalone called desktop or server based. GIS can run on both of them.

ii) **Software:** GIS Software provides tools and functions to input and store spatial data or geographic data. It provides tools to perform the geographic query, run analysis model and display geographic data in the map form. GIS software uses Relation Database Management System (RDBMS) to store the geographic data. Software talks with the database to perform geographic queries.

iii) **Data:** Data are the fuel for the GIS and the most important and expensive component. Geographic data are the combination of physical features and it's information which is stored in the tables. These tables are maintained by the RDBMS. The process of capturing the geographic data are called digitisation which is the most tedious job. It is the process of converting scanned hardcopy maps into the digital format. Digitisation is done by tracing the lines along the geographic features, for example to capture the customer building it should be traced around the building on the image.

iv) **People:** People are the user of the GIS system. People use all above three components to run a GIS system. Today's computer are fast and user friendly which makes it easy to perform geographic queries, analysis and displaying maps. Today everybody uses GIS to perform their daily job.

## GIS Data models

There are two major GIS data models:

i) **Vector Model**: There are three types of vector data, points, lines and polygons. The fundamental primitive of the vector model is a point, and the various objects are created by connecting the points (x,y) with straight lines, but some systems allow the points to be connected using arcs of circles. This type is used to store data which have discrete boundaries like country borders, land parcels and roads.

ii) **Raster Model**: Raster data store information about features in a cell based manner. Satellite images, Photogrammetry and scanned maps are all raster based data. Raster model is used to store data which vary continuously as in aerial photography, a satellite image or elevation values (DEM- Digital Elevation Model).

## The Impact of GIS on Physical Distribution

After development of GIS over the years, most businesses have considered it as a value added element in improving the distribution aspect in logistics, through using the software in route planning, optimisation, modelling, network maintenance, fleet management and delivery assessment. Nevertheless, the GIS technology is considered throughout all the value chain to support inbound and outbound logistics by adding value to administration, human resources, technology development, procurement, sales and marketing, services and operations. Now, most businesses have considered the GIS as a modern tool to improve logistics distribution and solve problems. The following are some examples:

i) GIS is a critical tool for logistics professionals because it manages massive amounts of location-based data to produce information that helps executives to make better choices.
ii) The GIS technology is used in diverse areas as utilities planning and retail location planning, as it a superb modern technology to face new challenges in logistics distribution.
iii) The GIS is a spatial component includes all types of data that can be used to track and manage both resources and processes. The resources could include distribution points for trucks or ships; finished goods, whether in a stationary location or vehicle; and raw materials or component inventory. Processes, on the other hand, may include assembly, transportation routing, merge in transit, demand or supply forecasting, or manufacturing lines.
iv) The GIS is able to integrate information and look at the potential impact of product location, transportation delays at distribution hubs, geopolitical turmoil on transit routes, availability and arrival time of materials, and projected requirements of customers.
v) The GIS is an important tool can be effectively used in decision making. By using the GIS technology, the fleet managers and executives are able to look at numbers in reports, charts, and graphs, thus they might adjust operations based on what they learned or use the derived information in formulating distribution plans.
vi) The GIS can also provide the opportunity to represent the information visually, allowing the decision maker to visualise a complete company profile to include manufacturer, office and warehouse locations, employee, client, customers, distributors and supplier locations.

vii) The GIS is able to uncover any issues might have occurred in logistics through its Data-driven analysis tool in a way to enhance processes and avoid errors.
viii) The GIS through its video surveillance is a valuable resource in making a confident assessment of possible corn-stover logistics distribution solutions.
ix) The GIS system can be integrated with other systems in an organisation via ERP, which can improve the processes and reduce the chances of introducing errors by manual intervention.
x) By using the GPS data-logging system along with GIS software, the fleet manager is able to conduct an accurate assessment of machinery performance by retrieving parameters such as position, time, speed, and fuel consumption.

**Advantages of GIS**

The following are the main benefits of GIS:

i) Integrating geographic information for display and analysis within the framework of a single consistent system.
ii) Allowing manipulation and display of geographic knowledge in new and exciting ways.
iii) Automating geographic information and transferring them from paper to digital format.
iv) Linking location and attributes of feature(s) within the framework of one system.
v) Providing the ability to manipulate and analyse geographic information in ways that are not possible manually.
vi) Automation of map making, production and updating.
vii) Providing a unified database that can be accessed by more than one department or agency.
viii) Storing geographic information in coincident and continuous layers.

ix) Providing an opportunity to open a larger number of income streams to those looking to sell their services and data to a broader online audience, anywhere round the World.
x) Playing an important role in cost savings through its greater efficiency, e.g. labor savings from automating or improving a workflow.
xi) It is a means for better decision making about location.
xii) It has the ability for improving communication through using GIS-based maps and visualisations which are greatly assisted in understanding situations and story telling.
xiii) It is a better means for geographic information recordkeeping.

**Disadvantages of GIS**

i) GIS technology might be considered as expensive software.
ii) It as well requires an enormous data input amount that is needed to be practical for some other tasks and so the more data that is to put in.
iii) Since the earth is round and so there would be a geographic error that will increase as you get in a larger scale.
iv) GIS layers might lead to some costly mistakes once the property agents are to interpret the GIS map or the design of the engineer around the utility lines of the GIS.
v) There might be failures in initiating or initiating an additional effort in order to fully implement the GIS but there might be large benefits to anticipate as well.

## 6) Distribution Requirements Planning (DRP)

The last technology in this chapter is Distribution Requirements Planning (DRP). This is another important technology in success of distribution by enhancing the overall effectiveness and efficiency of the logistics system. The DRP system, which has become a standard approach for controlling and planning distribution activities since the 1980s, is the process in which goods are delivered in a more efficient manner. These include considering the aspects of establishing a good, the quantity of the good, and the direct location that it is needed to arrive at in a given time. Furthermore, the system is a time-phased replenishment approach in which inventory status is reviewed and new shipment plans are generated periodically. The concepts and logic used are similar to those used in Material Requirements Planning (MRP). However, the approach is applied towards the distribution of goods separated geographically rather than the flow of parts within one manufacturing facility. The DPR is considered a general framework for planning and managing inventory in distribution networks, and applied to complex distribution networks with thousands of unique stock-keeping units and hundreds of stocking locations. Since DRP works similar to MRP (which already explained in previous chapter) then we decided to mention the main role and advantages of the technology that related to Distribution Management.

**Main roles of DRP**

The following are the main roles of DRP in distribution logistics

i) Supports time-phased requirements planning for single and multi-facility distribution organisations.
ii) Manages the flow of materials between firms, warehouses, distribution centers.
iii) Calculates inventory requirements over time and automatically generates online purchase requisitions.
iv) Provides data for managing the distribution facility and the database for consistent communications with the customers and the rest of the company.
v) Uses customer demand in the form of sales orders and forecasts.
vi) Considers current inventory levels and associated replenishment methodologies.
vii) Evaluates supplier information such as lead times and purchase order requirements.

viii) Calculates expected inventory levels over time and generates purchase requisitions.
ix) Support the organisation's inventory plan.
x) Supports online review of purchase requisitions and automatic generation of purchase orders and intercompany transfers.
xi) Supports centralised purchasing, facility-based purchases, or a combination of these methodologies.

**Advantages of DRP**

i) Balance customer service and the cost of distribution in complex networks.
ii) Develop distribution plans faster and more accurately, reconciled with material availability.
iii) Identify the operational and financial impact of demand and supply misalignments early.
iv) Make faster, more cost-effective decisions through cross-functional adaptive collaboration.
v) Make smart trade-offs to balance supply, demand and inventory risks.
vi) Push or pull supply distribution according to demand signals to reflect network limitations.
vii) One of the main benefits of using DRP is simply that inventory is managed in a smart way, that ensures maximum efficiency, with adequate levels of stock. Inventory tends to be an asset that is expensive in terms of how much resources it requires, so managing the inventory is important.
viii) It can keep inventory levels at a satisfactory level, sounds straightforward.
ix) It can create a system where the supply plan is generated automatically based on the forecasted demand, means that production and flow will be uninterrupted, which keeps customers happy through time scales and deadlines being met.
x) It helps to reduce procurement costs and procedures. Because stock levels are simply kept ticking over, those involved in the

process of buying stock will not have to deal with as many problems (i.e. stock not being available etc.), their time is freed up and the procurement system as a whole becomes more efficient.

xi) It keeps the stock at the lowest, and the available stocks are only stored to meet the business requirements.

xii) It provides savings by better aggregation of transportation and dispatching. In other words, DRP provides the data to accurately say when availability will be improved and delivery can be expected.

# PART FOUR
# Technology Applications in Bahrain

# CHAPTER 10

# Technologies in Material Management

## 1) Introduction

This chapter is about the Kingdom of Bahrain and describes the existing technologies being used in Materials Management by both the public and private sectors. This chapter was included in this book in order to comprehend how businesses in Bahrain are updated with the new technologies in the field of logistics, and to find out gaps that could be studied by researchers and practitioners to bridge it in order to improve operations and processes within Materials Management. Other reasons to consider Bahrain as the part of this book are:

i) Bahrain is the author's country of origin, therefore, focusing on technologies in Material Management can become more realistic and feasible.
ii) Bahrain is aiming to accomplish the vision 2030, and author's contribution to this accomplishment is through a study of the existing technologies being implemented by both the public and private sector related to Material Management, in order to make it successful eventually.

The chapter includes the country profile from the technology aspect, along with the latest technologies that have been adopted by the government for enhancing operations and processes in the public sector, and by prominent companies for an effective Material Management in the private sector.

## 2) Technology in Bahrain

Bahrain's strategy in the technology aspect is driven from an earlier plan called SISP (Strategic Information Systems Planning), which was proposed to develop open-standards based on the national IT infrastructure starting in 1993. Bahrain's regional leadership in Information Technology (IT) has been acknowledged by the World Economic Forum and INSEAD in the Global Information Technology Report 2007-2008. Bahrain is ranked $45^{th}$ global information and communications technology readiness index ahead of many countries in the region.

Traditionally, the government of Bahrain has been in the forefront of IT developments as compared to other countries in the region. It was perhaps the first country in the region that introduced computer in the government, and Bahrainis have proved to be an ideal testing ground for the introduction of new technologies in the region. Numerous ICT projects have taken place in the country in order to keep in line with any progress within the ICT field. As a result, the country has articulated an ambitious national economic development master plan called 'Economic vision 2030 and the National Economic Strategy', in which the ICT element forms the core amongst other elements in order to improve the standard of living through increased productivity and high-wage jobs. As part of the strategy, the government aims to achieve and accelerate its vision and strategy through the vision, and ICT plays a critical role in governance by situating technological innovation at the center of government discussions. ICT has also an impact in improving people's lives through information flows and communications. Additionally, ICT plays a very important role in improving the economic aspect for enhancing economic growth and income through improving productivity and quality of life. Bahrain as mentioned earlier has a highly advanced ICT infrastructure and the highest mobile and internet penetration rates in the region. The ICTs in Bahrain have been developed in recent years through the following features

i) Completely digitise its national and international phone switches.
ii) Embarking initiatives to improve the alignment between business and information technology by enhancing the ability of the government entity to better control IT-related changes in a manner that supports the overall business strategy.
iii) The government launched the e-Government portal to provide the best services and programmes at the highest levels for the business sectors and individuals.
iv) Introduce 3G and 4G high-speed download services.
v) Fully deregulate and liberalise its telecommunications market.
vi) First nationwide NGN (Next Generation Network) in the world.
vii) Cloud Computing Initiatives.
viii) Supporting projects to ensure alignment of IT projects with main strategic objectives of vision 2030.

Bahrain's vision is considered a tough challenge which aims to meet the following objectives:

i) Improve the skills of Bahrainis in all sectors.
ii) Focus on improvement in quality of services being provided by the government.
iii) Encourage Bahrainis to enter global markets through innovation and development.
iv) Make use of the extraordinary opportunities in the GCC countries.
v) Make a digital transformation in all areas of businesses in Bahrain.

## 3) Realising the Importance of Supply Chain by the Government

### a) **Bahrain Logistics Zone (BLZ)**

Strategic location in the heart of the Arabian Gulf region is considered a rarity in terms of establishing a business and transportation hub for the Northern Gulf, and a gateway to the Middle East. Such as the most notable location advantage of Bahrain has offered the country an efficient access to a growing regional customer base. To this end, the Kingdom is investing over US$32 billion in key infrastructure projects that will support the continued development of the logistics sector. One of the examples is Bahrain Logistics Zone (BLZ), which occupies 1 Km² of prime real estate close by Khalifa Bin Salman Port Bahrain's newest and biggest port. The zone offers customised and tailor-made services, ensuring there is a package to suit the operational needs of every business. Furthermore, the BLZ is thought the ideal location for many businesses by offering unique benefits that come with the zone's integrated world-class services. The BLZ offers the most cost-effective base for doing business in the Northern Gulf. The main benefits of choosing Bahrain Logistics Zone include:

i) 100% foreign company ownership is allowed.
ii) Multimodal access by land, sea and air.
iii) Flexible plot options with a range of plot sizes starting from 3000 m² onwards.
iv) Purpose-Built Infrastructure with an ample road network and plot entry allowances, designed for lorry traffic needs.
v) Dedicated 24-hour customs services.
vi) Basic services such as facilities management and special waste management.

Further, the BLZ offers customised and tailor-made services, ensuring there is a package to suit the operational needs of every business. With dedicated customs, adaptable plot options and easy on-site government services, organisations will be equipped with everything needed to facilitate the movement of goods from the BLZ to the Gulf and beyond. Land plots available for lease start from 4,000 square metres. Whatever the choice of land size or usage, all plots are equipped with full utilities, security and other management services. In addition to packaging, repackaging, logistics, export and re-export services, the BLZ provides dedicated and experienced account managers, who offer a bespoke, personalised service and ensure the delivery of seamless end-to-end solutions. In coordination with related government and legal entities, these account managers can assist with licensing, registration, visa processing and other related procedures to establish operations in Bahrain. For such outstanding services, the logistics performance was reported at 3.314, ranking overall (1=low to 5=high), in Logistics performance index according to the World Bank collection of development indicators in 2016, and the transport and logistics contribution to the GDP increased from three per cent in 2013 to seven per cent in 2015, demonstrating the growth potential of this important sector. In addition to the BLZ, there are other regional centre for logistics, such as the Sitra Causeway, Umm Al Hassam interchange, Khalifa Bin Salman Port and the upgrade of Bahrain Int'l Airport, which is estimated to increase the air cargo volumes to 1 million metric tonnes per year once the project is completed in 2020, and all these projects indicate the government of Bahrain is giving attention to logistics as the main source of government income and economically prosperous.

b) **e-Government Portal (www.Bahrain.bh)**

Bahrain launched its first e-Government strategy in 2007, by building the national portal (www.bahrain.bh) as a one-stop shop to facilitate the diffusion of various e-Services, expanding the integration of e-Government infrastructures and initiatives towards maturity, as well as promoting the e-Transformation process. The reasons for developing the e-Government portal are various, which are summurised below:

i) **Political:** The e-Government portal has been developed to let people interact with the government through different levels, and to help citizens and other residents connect to the government for having government services in a short way through electronic means, 'G2C'.

ii) **Business:** The e-Government portal is developed to enable businesses (private sector) process their activities in a healthy environment and easier, with improvements to the way in which businesses are registered, and it provides access to multiple private sector services required by startups. 'B2B'.

iii) **Economic:** The e-Government portal is a tool in economic development, reforms and also makes government more effective. The portal is developed to offer tremendous services such as e-Procurement, e-Banking, e-Commerce or an online trading exchange through e-Transactions between all sectors in the society, which is known as 'G2B'.

iv) **Administrative:** The e-Government portal is developed to provide interdepartmental services and structure in order to improve information and merge related services, and it's called 'G2G'. The main advantage of such an action is to simplify and facilitate easy delivery of services without bureaucracy, and to reduce corruption within the society.

v) **Employees:** The e-Government portal is developed to boost knowledge through sharing resources with others regardless of differences in experience and background, which is known as 'G2E'. The objectives behind such attention was because of the frenzy of the internet putting the entire public administration under the umbrella of the internet.

## The e-Government Portal and Procurement

One of the main reasons for the e-Government portal is to to deliver electronic procurement to enhance transparency, establish an open marketplace for procurement needs, and support the introduction of procurement reforms to better manage and monitor public procurement activities. The initiative is able to create a number of new options and methods for supporting the procurement processes of governments and for embracing the efficiencies and savings that can be realised through such the e-Service. The government of Bahrain uses the portal for conducting the procurement processes and relationships with suppliers for the procurement of works, goods, and consultancy services required by all ministries under its umbrella "public sector". Furthermore, the portal is used by all ministries to conduct all procurement process cycle, fully electronic and integrated in order to achieve value-for-money, contributing to socio-economic development, and leveraging information and communication technologies in the country. The main contribution of the e-Government portal to Procurement can be summarised as follows:

i) Aims to improve the government performance, streamline the procedures pertaining to tender and government procurement.
ii) The portal enables purchasing entities from the government sector in its first phase to prepare the purchase order, submit tender documentations.
iii) The portal allows suppliers and contractors from the private sector to purchase tender documentations through the said system and the procedures will be completed as usual.
iv) The portal covers the full cycle of the e-Tendering process, starting from submitting applications of offers, announcing tender invitations for review of the willing bidders to accept and depositing offers submitted by bidders with confidentiality and protection.
v) The portal offers transparency and clear evaluation of the offers, and finally announcing the awarded company, all being done online.
vi) Through the e-Government portal, the government could save millions of Bahraini Dinars per annum for the Government of Bahrain by unifying the purchasing of frequent goods which are an integral part of government operations.

vii) The portal contains information on laws, rules, regulations and tender-related data, in addition to news and activities.

viii) The portal provides features of live broadcasts for the opening tenders.

## 4) Companies and Organizations in Bahrain (Private Sectors)

Bahrain's track record as a business-friendly environment is decades-old. Home to the Gulf's long established financial services sector, the Kingdom has always stood out as one of the region's most liberal economies with 100% foreign ownership allowed for most businesses, and the government always identifies the motivation factors for investors to start a business in the kingdom and engage them in a collaborative manner to work towards gaining a competitive advantage in the market. Prince Salman, who is leading the economic development in the kingdom, outlined three economic and socio-economic priorities that are being pursued across government, comprising an increased role for the private sector in driving economic growth; supporting greater innovation and competitiveness, which represent the key pillars economic roadmap towards the Bahrain's Vision 2030. Since logistics play an important role to attract companies to start up their businesses in the kingdom, the government has launched many projects in this regard. For example, the government has pledged $694 million in ongoing investment in infrastructure projects and roads, which included the reconstruction of the Sitra crossing which was completed in 2011, and the Shaikh Khalifa Bin Salman Highway, adjacent to the Mina Salman intersection project. For such a healthy business environment, the Logistics sector in the kingdom has attracted a number of leading international companies to establish headquarters in Bahrain in order to access the neighbouring GCC economies, including DHL, Aramex, KWE and Agility.

## Technology Applications in Material Management by the Private Sectors

In this section, we list the current technologies being implemented by the private sectors in Bahrain, which enable them to coordinate the activities to manage the three main pillars of material management (warehousing, procurement, and distribution).

a) **Warehousing**

The most warehousing technologies currently used by companies in Bahrain are as follows:

i) **Enterprise Resource Planning - ERP**

The ERP system has been deployed by some large companies in Bahrain like the Aluminum company (ALBA) and Bahrain Telecommunications company (Batelco) since 2001. Then, many companies have been motivated to adopt it after improved outcomes and successful implementation by the two mentioned companies. However, one of the biggest disadvantages of ERP implementation in Bahrain is lack of integration between WMS and ERP. While ERP is considered an innovative means that is able to boost overall efficiency in warehousing, but if it is not integrated with the WMS the business might lose out on massive potential, and hence difficult for a company to achieve the agility to become more competitive in the market. Moreover, it is important to mention that the WMS is totally not considered by any business in Bahrain, and that is one of the reasons companies are still facing many problems of stock control and can't achieve the overall efficiency in warehousing. The main role of the standalone ERP technology that is currently used by companies in Bahrain is to integrate the key warehousing processes with other functions as explained in detail in the previous chapters. The following are the main functions of ERP being implemented and utilised by most businesses in Bahrain, according to our observation, especially the largest ones are:

i) Controlling bin locations through visibility of product location by batch and serial number, updated in real time manually, i.e. without using barcode, RFID or voice capture devices.
ii) Tracing items cradle to grave from receipt, through to storage, production, dispatch and return in real time, using manual input.
iii) Issuing materials defined by the batch/serial number to specific work orders or users using manual input process.
iv) ERP is used to receive goods and document all good receiving related transactions, such as Goods Receiving Notes (GRN) and gate passes, with automated workflows.
v) Easily identify and reduce excess stock levels, and reduce carrying costs of unnecessary inventory, along with locating obsolete inventory to remove from a warehouse.
vi) Analyses historical demand for every SKU across a warehouse.
vii) Implementing MRP and Forecasting.
viii) Unified methodology for monitoring the implementation of warehouse orders, including alarms for delays, procurement and production are the most concern in this aspect.
ix) Used as a platform that integrates various landscapes of manufacturing such as planning, purchase, sales, engineering, production, finance, Accounting, and delivery. The process through ERP work based on a common database, and it regulates the flow of information among them.
x) The system is used to classify products as categories ABC based on how often they are picked by a warehouse.
xi) The system is used as the reporting mechanism of inventory within the store and warehouse locations, to

help the warehouse for planning their future production schedules.

xii) The system is used in stock counts through its accuracy feature by measuring and comparing the actual physical stock of the product in the warehouse with the stock of that product in the system of record. Measuring stock accuracy involves periodic physical counts, reconciling physical and systems stock and making necessary adjustments to the system's stock.

ii) **Reach Forklift Trucks**

This is another technology is used in warehousing by some companies in Bahrain. The reach forklift truck is considered a modern styling with state-of-the-art engineering through its class-leading residual capacities, high efficiency, superior ergonomics and low maintenance needs. Moreover, the truck is almost used in narrow aisle applications by companies whose warehouses are large with high-storage racking system. The key ability of the reach truck is that it can extend its forks beyond the compartment and reach into warehouse racks in a way that standard forklifts cannot. It also features an open compartment that allows the operator to have greater visibility.

iii) **Very Narrow Aisles Truck - VNA**

VNA is used by a few companies that prefer the electric engine, battery and operator compartment to counterweight heavy loads, in order to ensure extremely fast pallet handling with much higher productivity and handling times than reach trucks or counterbalance forklifts. We noticed the main reasons for utilising the VNA truck is to maximise space utilisation by making the aisles as small as possible, and Deeko Bahrain one of the manufacturing companies relies heavily on such the truck. Also, the truck is able to access each pallet position easily, and it is specialised in increasing the number of pallet places in high density storage and it is equipped with articulated steering for decreasing the turning space while changing aisle. The VNA is the best way to implement the first in first out method.

iv) **Automated Conveyor System**

Material handling conveyor is one of essential components used by some large companies in warehousing in Bahrain, especially manufacturing companies. The technology is an important component for them to add speed, efficiency and safety to warehousing operations. The reasons for using it are:

i) To enable faster transport of products from production to storage to the staging area, especially in warehouse configurations with a great deal of distance to be covered, that are nonlinear, and/or involve vertical movement.

ii) Along with greater speed, which can dramatically increase throughput, automated conveyor systems move products more smoothly, reducing or eliminating product and pallet damage. Product damage, especially when undetected, results in returns, reshipments, and ultimately, damage to customer relationships, or even lost customers.

iii) In addition to speed and damage reduction, automated warehouse conveyor systems enable personnel to replace physically demanding, repetitive labor with tasks that add more efficiency to warehouse and shipping operations, and last but definitely not least, help to reduce injury, absence from work and downtime.

iv) To reduce labour costs, since the movements of goods in a warehouse take place without or less the use of manual labor.

### v) Overhead Crane

This technology is mainly used by Aluminum Bahrain (ALBA), as it is one of the largest companies in manufacturing finished aluminum products. The technology is commonly used for lifting and hoisting raw materials from the warehouses to various plants. The component consists of parallel runways with a traveling bridge spanning the gap. The warehouse uses the component to pour raw materials into plants, using cranes which travelling along the bridge. The overhead cranes are used by the company at every step of the manufacturing process, until it leaves a factory as a finished product.

### b) Procurement

Most companies in Bahrain use an ERP software system to run the entire process of procurement. The procurement department is connected with other departments within an organisation, using a central database that links different departments of an organisation, and that strives to create a partnership with each of the organization's departments. One of good example in this aspect is Bahrain Telecommunication Company (Batelco). The company has implemented the SAP ERP software, which consists of several important components and modules, including the logistics module. The module covers business functions such as procurement and inventory management. The procurement process through the module is done for materials and services that are required for business from both local and overseas suppliers. However, according to our observation, we haven't noticed any procurement system within companies in Bahrain is connected with suppliers either internally or externally through the ERP system, which is known as B2B, in which a company's procurement can deal with suppliers through an integrated system for making purchase orders and other related activities, and that could due to some reasons related to security issues, or lack of experience in such business activities. The main activities of the ERP system within the procurement in most companies in Bahrain are:

i)  **Purchase Requisition:** If a product is required through the production planning process or warehouses a purchase requisition is generated either by a user or buyer within the procurement department.

ii) **Request for Quotation (RFQ)**: If the product cannot be bought from a contracted supplier or the price and quality is not acceptable; the procurement uses ERP to send new or existing suppliers an RFQ. Also, If a requisition is raised for an item that has not been purchased before, a request for quotation is created and sent to existing or new vendors, and that is done by the procurement department.

iii) **Purchase Order**: The procurement department issues POs to the vendor, using the ERP system.

iv) **Contracts:** The procurement uses the system to create a contract as well, where the vendor agrees to send a certain quantity of the items over a certain period.

v) **Scheduling Agreement**: The system is used to schedule delivery intervals over a period with a supplier.

vi) **Control Budget:** The system is used to control the procurement department's budget.

vii) **Forecasting:** The system is used to forecast and review purchases in comparison with both production and warehouse records.

viii) **Vendor Master Data:** The procurement uses the system to evaluate vendor performance.

ix) **Expediting:** The system is used to expedite POs with suppliers.

c) **Distribution**

Companies in Bahrain run the distribution in logistics either as part of warehousing roles or through a third party company. The roles of the distribution are mainly focused on handling outbound, inbound and internal transportation. The main activities of the distribution roles in most companies in Bahrain are:

i) Deliver and track goods from its source to their destination e.g. wholesalers and small, overseas customers, or suppliers for returned goods.
ii) Identifies every cost element that is associated with the movement of products.
iii) Facilitates the transportation methods for an effective delivery.
iv) Facilitates shipments through air and sea freight.
v) Provides a very organised shipping system that assists in time and cost saving.
vi) Provides order fulfillment flexibility using unparalleled resources for multi and single site operations.
vii) Takes care of packaging, especially for those being exported abroad.

**Technologies used by Distribution in Bahrain**

The main technologies used by companies in Bahrain are:

i) **GPS:** This is one of the key technology used by most companies in Bahrain to gain greater vehicle visibility and maximize productivity in tracking of goods, vehicles, and drivers. Also, the technology is used to keep updating customers about their locations accordingly. There are many GPS providers in Bahrain offer latest GPS technologies with the following features:

    i) Quick and easy installation in vehicles.
    ii) Working with high secure web portal.
    iii) Track vehicles from any PC or Laptop connected to the internet.
    iv) Minimal IT involvement or integration.
    v) Receive and display alarms (SOS, speeding, ignition, antenna-cut, power-cut, etc.).
    vi) Increased driver productivity.
    vii) Automated driver hours and mileage reports.
    viii) Automated arrival and departure times.

ix) The elimination of trip sheets.
x) Improved fuel consumption.
xi) Wireless connectivity.
xii) Numerous detailed automated reports.
xiii) GPS location time-stamped.
xiv) Support Google Maps and Google Earth.
xv) Proof of driver and vehicle activity.

ii) **Inventory Control Handheld Device:** This technology is used by some companies in Bahrain to control inventory being delivered to customers. The device is used by drivers who are responsible to deliver goods to customers, which enables them to control stock being transported in order to ensure accuracy, reliability, and productivity of any shipment. Also, the device helps drivers to immediately check finished goods and provide customers the most current inventory and price information. One of companies in Bahrain that is using the device is Pepsi-Cola, and according to our observation, they could ensure a good distribution controlled process through such the device. The handheld device works on a real-time base and has the following features from the distribution aspects:

i) It reduces wait time for customers while employees manually check stock shelves for items or look up information on a PC.
ii) It reduces data entry time.
iii) Improves Inventory Accuracy.
iv) It increases productivity through streamlining employee utilisation.
v) It enables full process automation.
vi) It enables to identify inventory inaccuracies faster.
vii) It helps to optimise replenishment and fulfillment plans, and decreasing the carrying costs of excess materials.
viii) It helps to improve the quality of supply chain collaboration, and optimises warehouse and supply site scheduling activities and resources.

ix) It helps drivers to significantly reducing dependence on adequate space and allowing them to access the system wherever they are.

iii) **Packaging Technologies**

Packaging is the role which is normally handled by the distribution section in most companies in Bahrain, either through own business or by a third party. Different methods and techniques are used to achieve an effective and efficient packaging with them as it is considered as an aspect of a company's profitability and reputation, the secondary and tertiary are the essential parts of shipment that are covered by physical distribution. The following are the packaging technologies used by companies in Bahrain:

a) **Pallet Wrapping Machines:** This technology is used to eliminate the need to manually wrap pallets. The machine entails the use of stretch cling Film. The stretch film is stretched to the required percentage and wrapped around the product/pallet. The film clings to the previous layer of Film. The cycle keeps on repeating till the entire product/pallet is covered with a film. The machines are mainly used by manufacturing companies such as (ALBA, Pepsi Cola, Deeko Bahrain, and others). The machine is used for various products such as: Moulded Plastic Furniture, Cartons, Textile rolls, Cans, Aluminum and plastic containers, and more. The main advantages of this machine are:

　i) Protection from dust, moisture and foreign particles.
　ii) Protection from pilferage.
　iii) Gives stability to the pack.
　iv) Scratch resistance.
　v) Handling becomes Easy and Fast.
　vi) Economical mode of packing.

vii) Reduction in costs of labours.
viii) Reduction in costs of resources.

b) **Palletiser:** The machine provides an automatic means for stacking cases or other package type of goods or products onto a pallet for efficient transportation and storage. The act of placing cases or other package types upon pallets for handling, moving or storage. The main advantages of the machines are:

i) Stacks better, more accurate quality load than manual labor.
ii) Replaces costs of manual labor and less workmen's compensation claims.
iii) Error reduction and increased production speeds.
iv) Decrease in product damage and shipping costs.
v) Provides standardisation.

c) **Carton Packing Machines:** The machine is used for sealing cartons automatically, and it is mainly used in factories to be ready for physical distribution once they get out of the plants. The machine is used in all industries that use cardboard boxes for packaging and shipping goods. Common industries in Bahrain which use carton sealing machines include: Food and beverage, production, Perishable food goods, Textiles, Consumer goods, etc. The main advantages of the machine are:

i) Improved production efficiency.
ii) Eliminate bottlenecks caused by handheld tape dispensers jamming and running out.
iii) Ensure a consistent and professional result for every carton.
iv) Reduces labours.
v) Ensure standardisation.

## 5) Other Technologies used by the Private Sector in Bahrain

a) **Point of Sale:** All supermarkets and some restaurants in Bahrain use the POS system as the electronic equipment performing the sales transaction and processing, either through cash, debit cards, or e-Wallet. Moreover, the system is used for other essential tasks such as inventory management, sales reporting, customer management, marketing initiates, and etc. The system is a combination of software and hardware that allows merchants to take transactions and simplify key day-to-day business operations:

   i) **Point of Sale Software**: software that enable a retailer to check out a transaction (and may also run the store's operations).
   ii) **Point of Sale Hardware**: computer equipment that facilitates the checkout process (electronic cash register, cash drawer, receipt printer, bar code scanner, tag printer).
   iii) **Point of Sale Terminal:** payment hardware terminal that enables a retailer to process and accept payment from credit/debit cards

The following are the main advantages of POS as reported by businesses in Bahrain:

   i) **Sales History:** The system can easily look up past transactions and discover which product is stuck on the shelf for weeks as well as which products are selling the most, which will make inventory management that easier.
   ii) **Inventory Management**: The system is the best tool to accurately control inventory. The system is able to control stock of each product according to its name, brand, supplier, supplier code, SKU, color and a multitude of other categorisations.
   iii) **Sales in Real-Time:** The system allows businesses to have all transactions in real-time, including reports. The

POS system also allows to customise reports according to business requirements.

iv) **Saving Time and Money**: The system is the best method to make any transactions faster than manual cash registers. Everything from reading the product barcode to deduct the price from a debit card or pay case can save a lot of time, thus making the customer and cashier happier. The system also allows business to use add-on devices to help make things quicker and make the sale faster, such as electronic cash drawers, barcode scanners, credit card readers and receipt or invoice printers. No more writing it down on a piece of paper or summing it up on the calculator.

v) **POS Software:** The system could be integrated with other modules, such as accounting modules, including general ledger, accounts payable and receivable, among other features.

vi) **Financial Control**: The system allows to abolish the "put it on the cuff" way of selling, and helps businesses to manage sales accurately. Therefore, the accounts receivable and payable may be all accounted for and hence there will not be any cash breakages or shortages anymore.

vii) **Management and ERP Systems Connectivity**: The system can be connected to other terminals and to a back-office server, and it can also be plugged into an ERP system in order to manage everything: sales, inventory, orders, accounts payable and receivable, etc.

b) **The Cargo Carousel System (CCS):** This system is used by cargo companies in Bahrain. The system is a frame configured to fit within a shipping container that allows deliveries and pickups at the same time without ever leaving the dock whether transporting upstream or down. This can eliminate empty backhauls by combining the supply and reverse chains into one, doubling potential revenue streams with an import/export mechanism at both ends of the value chains. Double-stacking pallets will often crush products underneath, but the CCS eliminates partial loads by utilising the entire available space to further increase revenues while protecting merchandise far better than plastic wrap. The main advantages of CCS are:

i) Increase picking speeds and order accuracy.
ii) Improve operator productivity.
iii) Shorten pick time through reduced walk distance.
iv) Improve inventory record accuracy.
v) Easily enter additional SKU's for growing operations.
vi) Improve customer service and response time.
vii) Reduce shrinkage and product damage.
viii) Reduce or eliminate unnecessary paperwork.

**Case Study**

"The Role of Business Intelligence (BI) in Enhancing the Competitive Advantages of Organisation in Bahrain".

*Published by Ali Kamali (the author of this book), Master thesis, Arabain Gulf University, Bahrain, 2010.*

The researcher conducted his study to evaluate the business intelligent systems and their impact on the company's competitive advantage, using the implementation of SAP system in Logistics and Finance in Bahrain Telecommunications Co. (Batelco) as a case study. The researcher used both qualitative and quantitative methods in explaining the newly implemented system in Batelco. The methodology used to investigate the usability of SAP in improving the business in Logistics and Finance. The survey was conducted and included three main tiers through questionnaires that relevant to each tier. The groups included senior management, users, including managers of departments and finally some of the staff specialists in the Information Systems who deal with the technical aspect of BI.

The results obtained by the researcher as follows:

i) The results indicated that the company was able to be more powerful than ever before, and its profits exceeded as an indication of its strong position in the current telecommunications market.
ii) The results that achieved through the case study showed that the knowledge systems are the main priority for different management levels since 1996.
iii) All three management levels that included in the survey agreed the importance of BI in their respective levels and areas. In the second model, the results also showed that the company could effectively achieve its goals when the productivity of the SAP system tested and compared to the legacy system, and found the KPIs could meet the company's objectives.
iv) The results indicated that BI and Knowledge Management were the core elements to keep the company ahead.
v) Although the survey indicated Batelco was successfully dealing with BI, however, the company still needed to set an aggressive strategy for more accelerating the benefits of BI, so that to give more real value to the system. This point arose because there were many users yet not understanding the real value of BI on improving the business.

# APPENDIX

**Bibliography**

BARNARD, C (1938). *The Function of the Executive*. Massachusetts, Harvard University Press.

BARNARD, C. (1938). *The function of the executive*. ISBN 067432-800-0 ed., Boston, Harvard University Press.

BARONSON, J. (1970). Technology Transfer through the International Firms. *American Economic Review Papers and Proceedings*, 435-440.

BURGELMAN, R. A., Maidique, M. A., & Wheelwright, S. C. (1996). *Strategic Management of Technology and Innovation*. 2$^{nd}$ ed., Chicago, I. L, Irwin.

CAPGEMINI, Consulting (2017). *Global Supply Chain Control Towers*. [online]. HYPERLINK "https://www.capgemini.com/wp-content/uploads/2017/07/Global_Supply_Chain_Control_Towers.pdf" https://www.capgemini.com/wpcontent/uploads/2017/07/Global_Supply_Chain_Control_Towers.pdf

CHEN, Haozhe, MATTIODA, Daniel D and DAUGHERTY, Patricia J (2007). Firm-wide integration and firm performance. *International Journal of Logistics Management*, **18** (1), 5-21.

COSTA, A.C., & Bijlsma-Frankema, K (2007). Trust and control interrelations: New perspectives on the trust-control nexus. *Group & Organization Management*, **32** (4), 392-406.

DAVIS, F. (1989). Perceived Usefulness, Perceived Ease of Use, and User Acceptance of Information Technology. *International Journal of Man-Machine Studies*, **13** (3), 319-339.

DUNNING, J. H. (1994). Multinational Enterprises and the Global of Innovatory Capacity. *Research Policy*, **23**, 67-88.

EUROPLATFORMS (2004). *Logistics Centres Directions for Use.* [online]. HYPERLINK "http://www.unece.org" http://www.unece.org

FERRES, N., Connell, J., & Travaglione, A. (2004). Co-worker trust as a social catalyst for constructive employee attitudes. *Journal of Managerial Psychology*, **19** (6), 608-621.

HOWELL, J.P., & Costley, D.L. (2006). *Understanding behaviors for effective leadership.* 2nd ed., Upper Saddle River, NJ, Pearson Prentice Hall.

KOTTER, J. P., (2001). What leaders really do? *Harvard Business Review*, **79** (11), 85-96.

KUMAR, V., Kumar, U., & Persaud, A. (1999). Building Technological Capability through Importing Technology: The Case of Indonesian Manufacturing Industry. *Journal of Technology Transfer*, **24**, 81-96.

LAN, P., & Young, S. (1996). International Technology Transfer Examined at Technology Component Level: A Case Study in China. *Technovation*, **16** (6), 277-286.

LASTRES, S. A. (2011). Aligning through knowledge management. *Information Outlook*, **15** (4), 23-25.

LEUNG, J., & Kleiner, B.H. (2004). Effective management in the food industry. *Management Research News,* **27** (4/5), 72-81.

LEUNG, J., & Kleiner, B.H. (2004). Effective management in the food industry. *Management Research News,* **27** (4/5), 72-81.

SAWHNEY, R. (2013). Implementing labor flexibility: A missing link between acquired labor flexibility and plant performance. *Journal of Operations Management,* **31** (1/2), 98-108.

SKRBINA, David. (2015). *The Metaphysics of Technology.* New York, Routledge.

TAYLOR, W. F. (2008). *The Principles of Scientific Management.* Stilwell, KS, Digireads Publishing.

VOLTI, Rudi. (2009). *Society and Technological Change.* vol.7$^{th}$. New York, Worth Publishers.

WATERS, Donald (2003). *Logistics An introduction to supply chain logistics.* Palgrave Macmillan.

# INDEX

**Symbols**

1D barcodes  154
2D barcodes  155
3PL  27, 233, 237
3PLs  32
4PL  233, 237
4Pls  32
, GPS  242

**A**

ABC Analysis  68
ABC Technique  69
AGV  143
Air Transport  115
ANSI X12  167, 169
APICS  9
Application software  20
Auto ID  149, 161
Automated Guided Vehicle  146
automated guided vehicles  143
Automatic Identification Technology  148
Automatic Vehicle Location  246

**B**

B2A  203
B2B  31, 32, 188, 192, 269, 276
B2C  31, 188, 197
B2G  188, 203
Bahrain  263
Bahrain Logistics Zone  267
bandwidth  230
Bandwidth  191
Barcodes  154
BI  224
Big Data technology  223
Bills of Lading  117
block  217
Blockchain  135, 217
bloggers  189
BLZ  267
Bribery  97
Brick and Click  189
Bucket elevators  77
Budget control  187
Bulk Material Handling  76
Business Intelligence  224
Business-to-Administration  203
Business to Business  192
Business To Consumer  197
Buying a Warehouse  65, 66

**C**

C2C  206
C2M  192, 208
Cantilever Racking System  61

Cargo-Imp  167
Cargo-IMP  169
Carton Flow Racking System  63
Carton Packing Machines  281
Centralized Procurement  88
Chaster Bernard  6
Checksum  156
CII  9
CILT  9, 26
CIPS  9, 26
Cloud Computing  214
Cloud ERP  174
Code 39  157
Code 128  157
Coercion  97
commercial distribution  24
Communication  94, 166
Communication Networks  20
computer-based record  11
Computer Integrated Manufacturing  76
Connected Words  162
Consumer To Consumer  206
consumer to manufacturer  192
Continuous Speech  162
Control Tower  233
Conveyor belts  77
Conveyors  76
Conveyor System  275
coordination and collaboration  23
CPIM  9
CPU  19
Cranes  76
CRM  19
cryptocurrency exchanges  223
CSCMP  26
customer demand  9
Customer Relationship Management  19
Customer Service  104
Cyber Analytics  225
cybercriminals  216
cybersecurity  225

## D

D2C  209
dashtop mobile devices  212
Database  20
Database management system  20
Data characters  156
DBMS  20
DDMRP  73
Decentralized Procurement  88
Deliveries  81
Demand Management  24
Descriptive Analytics  225
Diagnostic Analytics  225
Digital marketing  188
Direct sellers  199
Direct To Customer  209
Dispatch of Orders  57
Distribution  99, 100, 232, 277
distribution channels  107
Distribution Channels  108
Distribution Function  102
distribution logistics  259
Distribution Management  24, 31
Distribution Manager  14
Distribution Requirements Planning  258
distribution system  101
Documents and Records  93
Drive-in Racking System  62
Drone System  144
DRP  258

## E

e2e  236
e-Auction  182
e-Business  188
e-Buy  188
e-Collaboration  182
e-Commerce  189, 205
Economic Order Quantity  70
EDI  167, 168

EDI Cloud  170
EDIFACT  167, 169
e-Fulfillment channels  239
e-Government  203
e-Government portal  266, 269
e-Government strategy  268
e-Informing  182
electronic data interchange  188
Electronic Data Interchange  170, 182
e- MRO  182
encryption  220
Engineered Systems  76
Enterprise Resource Planning  19, 171, 181, 182, 272
EOQ  70
e-Procurement  184, 186, 187, 203, 205
E-Procurement  180, 182
e-Reverse auctioning  182
ERP  19, 29, 32, 58, 135, 165, 166, 171, 172, 226, 276
e-Sale  188
e-Sourcing  182
ETA  235
e-Tendering  182, 270
ethics  96
Exchange Data Interchange  167
Exclusive distribution  110
Expediting  277
Extortion  98

# F

Favouritism  98
FCL  117
FEFO  149
Fetch Robotics  143
FIFO  149
finished goods  11, 30
Finished Goods  105
Flexible Manufacturing Systems  76
flow of materials  22
Forecasting  186, 277

Forklift  76
Frederick Winslow Taylor  5

# G

G2B  269
G2C  269
G2E  269
G2G  269
General Services Administration  205
Geographical Information System  252
GIS  252
Global Positioning System  245
Google Hangout  230
GPS  245, 278
Grain elevators  77
GRN  273

# H

Handheld Device  279
hard drive  19
hardware  188
Hardware  19, 254
hash  217
Hoppers,  77
HRD  18
hub-and-spoke  240

# I

ICT  203, 241, 265
IISCM  9
Illegal sourcing  98
Inaccurate inventory records  11
Inbound  145
Inbound Processing  135
Incoterms  119
Information and Communications Technology  241
Information flow  109
Information Management  58
Information system  19
Information Technology  18

Integrated Logistics Platform  27
Integrated SCM  25
Intelligent Transportation System  241
Intelligent Transport System  248
Intensive distribution  109
Interleaved 2 of 5  157
Internet of Things  175
inventory  30, 99
Inventory  8
Inventory Control  105, 186
Inventory Management  190
Inventory planning  10
IoT  175, 177
IS  19
ISM  26
Isolated Words  161
IT  18
ITS  241

## J

JIT  74
Jobless  223
Just In Time  74

## K

KNAPP  142
KPI  235

## L

LCL  117
Leading  17
Lead Logistics Partner  233
Lead Logistics Provider  236
Leasing  65
LIFO  149
LLP  233
local area network  161
Locus Robotics  142
logistics  27, 29, 99
Logistics  22, 30, 31, 32, 33

## M

Management  5
Material Handling Equipment  57, 75
Material Management  14, 19, 22, 79, 180, 232, 264
Material Requirement Planning  71
Material Requirements Planning  259
Materials Handling  12, 107
Materials Management  30, 99
Material Stock  8
memory  19
Mezzanines  77
MHE  75
MM  30, 79, 100
Mobile Agents  212
Mobile Cloud Services  214
Mobile Procurement  211
Modes of Transport  111
monitor  19
Motivating  18
M-Procurement  211
MRP  71, 171, 259
MRP II  73
Multimodal platforms  28

## N

Negotiating  92
Negotiation  230
non-stock materials  8

## O

ODETTE  167, 170
On-Premises ERP  174
Open Shuttle  142
Open-source robotics  142
operating systems  20
Order Processing  104
organisation  181
Organisation  15
organisational structure  33

Organizational Structures  33
organizations  100
OSs  20
Outbound  146
Outbound Processing  135
Overhead Crane  276

## P

P2P  32, 188, 206
Package labelling  116
packaging  99
Packaging  121, 280
Pallet Flow Racking System  63
Palletiser  281
Pallets  77
Pallet Wrapping Machines  280
PDM  19
People  21
PEOU  44
Perceived Usefulness  44
Percieved Easy of Use  44
personal digital assistant  212
person-to person  206
Peter F. Drucker  5
physical distribution  232
Physical flow  109
Planning  16
Point of Sale  282
Predictive Analytics  225
Prescriptive Analytics  225
private sectors  204
procurement  232, 269
Procurement  32, 78, 83, 87, 92, 95, 180, 190, 276
Procurement Manager  14
procurement's role  32
Product Data Management  19
Production Management  24
Programming software  20
PU  44
public sectors  204

Purchase Order  277
Purchasing  78
Purchasing Requests  81
Pure-Play  189
Push Back Rack System  62
Putaway  146

## Q

Quite Zone  156

## R

Racking Systems  60
Racks,  77
radio frequency  138
Rail Transport  114
Raster Model  255
raw materials  11, 30
RDBMS  254
Reach Forklift  274
Reclaimers  77
Relation Database Management System  254
Request for information  204
Request for Quotation  277
Request for quotations  204
Requests for proposals  204
return on investment  79
RF  138
RFI  204
RFID  133, 147, 148, 150, 151, 153, 154
RFPs  204
RFQ  277
RFQs  204
Road Transport  111
Robotics Technology  139
Robotic Systems  142
robotic technology  140
ROI  79

291

## S

SAP ERP  276
Scallog Robotic  143
Scheduling Agreement  277
SCIS  36
SCM  19, 36
Sea Transport  116
Security  216, 222
Selective distribution  110
Selective Racking System  61
Service Level Management  24
Service Oriented Architecture  213
Shelving  77
Silos  77
SKU's  146
SKUs  10, 149
Skype  230
smart transportation management  241
Software  19, 188, 254
spatial component  256
Spontaneous Speech  162
SQL  224
Stackers  77
Stacking frames  77
Staffing  17
Start and stop characters  156
STM  241
Stock Control  9
Stock Control Techniques  67
Stock counts  135
Stock Counts  147
Stock flow  186
Stock Planning  10
Stock Records  11
Storage drawers  77
Storage Equipment  77
Supplier  191
supplier evaluation  85
supplier relationship  24
supply chain  99
supply chain information systems  36
supply chain management  210
Supply Chain Management  19, 22
supply chains  234
Supply Management  24
Swisslog CarryPick Robot  143
Symbologies  155
System software  20

## T

TAM  43
TAM2  44
TAM3  44
Taylorism  5
Technologies in Warehousing  131
Technology  7, 35, 234
Technology Acceptance Model  43
Theory of Planned Behaviour  44
Theory of Reasoned Action  44
The purchasing function  14
The warehousing role  13
Third party logistics  27
Title flow  109
TPB  44
TRA  44
Traffic of influence  98
Transportation  110
transportation management systems  244
Transportation Mode  106
Transport Equipment  76

## U

Unimodal distribution area  28
Universal Product Code  157
usiness To Government  203

## V

validation  220
Variability  224
Vector Model  255
vehicles  242
velocity  224

Vendor Master Data  277
Vendor Rating  187
Very Narrow Aisles  274
Videoconferencing  230
Visibility  235
vision 2030  264
vloggers  189
Voice dependent models  162
Voice independent models  163
Voice Recognition Technology  161
VR  162, 166

# W

warehouse layout  64
Warehouse Layout  59

Warehouse Management System  138
Warehouse Manager  13
warehousing  99, 232
Warehousing  272
Waybill  116
Waybills  117
WIP  30
Wireless data applications  248
WMS  58, 134, 135, 138, 149, 161, 165, 166, 272

# X

X-dimension  156
XYZ Analysis  69

www.ingramcontent.com/pod-product-compliance
Lightning Source LLC
Chambersburg PA
CBHW021351210526
45463CB00001B/58